PHYSICS BY INQUIRY

An introduction to physics and the physical sciences

Volume II

Lillian C. McDermott
Professor of Physics

with

Peter S. Shaffer and Mark L. Rosenquist

and the

Physics Education Group
University of Washington

JOHN WILEY & SONS, INC.

Acquisitions Editor   Stewart Johnson

Marketing Manager   Catherine Faduska

Production Editor   Deborah Herbert

Designer   Harold Nolan

Manufacturing Coordinator Dorothy Sinclair

This book was set in 12 point Times Roman by the authors

ISBN 978-0-471-14441-0

10 9 8 7 6 5 4 3 2

Preface

Physics by Inquiry is a set of laboratory-based modules that provide a step-by-step introduction to physics and the physical sciences. Through in-depth study of simple physical systems and their interactions, students gain direct experience with the process of science. Starting from their own observations, students develop basic physical concepts, use and interpret different forms of scientific representations, and construct explanatory models with predictive capability. All the modules have been explicitly designed to develop scientific reasoning skills and to provide practice in relating scientific concepts, representations, and models to real world phenomena.

Physics by Inquiry is not meant to be passively read. The modules do not provide all the information and reasoning included in a conventional text. There are gaps that must be bridged by the student. The process of science cannot be learned by reading, listening, memorizing, or problem-solving. Effective learning requires active mental engagement.

Physics by Inquiry contains narrative, experiments and exercises, and supplementary problems. As the course progresses, student notebooks become an important resource.

- Narrative

   The narrative is double-spaced. It includes statements of fact, definitions, and examples of the kind of reasoning that is expected of students.

- Experiments and exercises

   Experiments and exercises are inset from the narrative. They should be done as they are encountered.

- Supplementary problems

   A collection of problems at the end of each module provides additional practice in applying physical concepts and scientific reasoning skills.

- Student notebooks

   Students maintain notebooks in which they record observations, do exercises and problems, and reflect on how their understanding is evolving. In this way, they create an indispensable reference that complements the text and serves as an individualized study guide.

Note to the instructor

Physics by Inquiry consists of three volumes. The first two are subtitled: *An introduction to physics and the physical sciences.* Volume I develops fundamental concepts and basic reasoning skills essential for the physical sciences. The material included in Volume II provides a foundation for the study of introductory physics. With the exception of *Electromagnets* and *Astronomy by Sight,* Volumes I and II can be used independently. Volume III, which is subtitled: *An introduction to physics,* introduces additional topics from the standard introductory course.

Physics by Inquiry has been designed for courses in which the primary emphasis is on discovering rather than on memorizing and in which teaching is by questioning rather than by telling. Such a course allows time for open-ended investigations, dialogues between the instructor and individual students, and small group discussions. A major goal is to help students think of physics not as an established body of knowledge, but rather as an active process of inquiry in which they can participate.

Physics by Inquiry is particularly appropriate for preparing preservice and inservice K–12 teachers to teach science as a process of inquiry. The modules can also be used to help underprepared students succeed in the mainstream science courses that are the gateway to majors in science, mathematics, engineering, and technology. For these student populations, as well as for those in the liberal arts, the curriculum helps establish a sound foundation for the building of scientific literacy.

Physics by Inquiry has an accompanying Instructor's Guide for college and university faculty. It has several purposes: to suggest how the materials can be most effectively used with different student populations, to help the instructor anticipate student difficulties, to provide information about the equipment, to describe the demonstrations referred to in the text, and to convey the purpose of unusual exercises and experiments.

Development of *Physics by Inquiry*

Physics by Inquiry is the product of an intensive, collaborative effort by the Physics Education Group in the Physics Department at the University of Washington. Directed by Lillian C. McDermott, the group includes faculty, research associates, and graduate students. Members of the group conduct in-depth investigations of student understanding through which they identify and analyze specific difficulties that students encounter in studying physics. This research has provided the foundation for the design of the instructional strategies that are incorporated in *Physics by Inquiry*.

Physics by Inquiry has been developed through an iterative, interactive process of research, curriculum development, and instruction. The participation of the Physics Education Group in the instructional program of the Physics Department has made it possible to design, test, and modify the curriculum in a continuous cycle on the basis of regular feedback from the classroom. In addition to monitoring the effectiveness with students at the University of Washington, the Physics Education Group has been able to draw on the experience of instructors at other institutions who have pilot-tested earlier editions of the modules. The extensive testing and subsequent revision that have been an integral part of the development process have helped ensure that *Physics by Inquiry* is well-matched to the students for whom it is intended.

Physics by Inquiry has evolved into its present form over a long period of time, during which the co-authors have shared intellectual leadership and editorial responsibility. In the early versions of the curriculum, Mark L. Rosenquist helped establish the instructional approach that underlies all of the modules. In all the later versions and in the additional modules that appear in this first published edition of *Physics by Inquiry*, Peter S. Shaffer has played a pivotal role.

Acknowledgments

Substantive contributions to the design, testing, and modification of the modules have been made by many members of the Physics Education Group, past and present:

Bradley Ambrose, Patricia Chastain, James Evans, Gregory Francis, Diane Grayson, Randal Harrington, Stephen Kanim, Christian Kautz, Pamela Kraus, Ronald Lawson, Michael Loverude, Martha Means, Tara O'Brien Pride, Graham Oberem, Brian Popp, Mark Somers, Richard Steinberg, David Trowbridge, Emily van Zee, Stamatis Vokos, Betty Windham, and Karen Wosilait.

Joan Valles deserves special recognition for her editorial assistance on all the modules. Helpful suggestions have been made by the peer instructors, preservice and inservice teachers, and undergraduate students in courses taught by the Physics Education Group.

Colleagues at other institutions who have contributed to the development of *Physics by Inquiry* include:

Davene T. Eyres (North Seattle Community College), T. Dean Gaily (University of Western Ontario), Fred M. Goldberg (San Diego State University), Eunsook Kim and Beth Thacker (The Ohio State University), Suzanne M. Lea (University of North Carolina at Greensboro), William Moore (University Preparatory Academy in Seattle), and Robert A. Morse (St. Albans School in Washington, D.C.).

The development of *Physics by Inquiry* has drawn on several published instructional materials. One important source has been *The Various Language* by Arnold Arons. Some of the precollege curricula developed with NSF support, such as *ESS, SCIS, IPS, PSSC Physics,* and *Project Physics,* have also been excellent resources.

Support by the National Science Foundation and the Physics Department has enabled the Physics Education Group to conduct the coordinated program of research, curriculum development, and instruction that has produced *Physics by Inquiry*. The encouragement of Clifford W. Mills at John Wiley & Sons, Inc., has also been deeply appreciated.

Lillian C. McDermott
August 1995

Contents

ELECTRIC CIRCUITS

ELECTROMAGNETS

LIGHT AND OPTICS

Part A: Plane mirrors and images

Part B: Lenses, curved mirrors, and images

KINEMATICS

Part A: Motion with constant speed

Part B: Motion with changing speed

Part C: Graphical representations of motion

Part D: Algebraic representations of motion

ASTRONOMY BY SIGHT: THE EARTH AND THE SOLAR SYSTEM

PHYSICS BY INQUIRY

Electric Circuits

In *Electric Circuits,* we examine the behavior of electric circuits consisting of batteries, bulbs, resistors, and wires. Using our observations as a basis, we construct a scientific model that we can use to predict and explain the behavior of simple resistive circuits.

Part A: Behavior of simple electric circuits

In Part A, we examine the brightness of bulbs that are connected to a battery in diffferent configurations. We introduce simple electric concepts that enable us to account for the relative brightness of the bulbs that we observe.

Section 1. Single-bulb circuits

We begin our study of electric circuits by connecting a battery and a bulb together and observing what happens. We investigate the conditions under which the bulb lights brightly, dimly, or not at all.

Experiment 1.1

A. Obtain one battery, one light bulb, and one wire. Connect these in as many ways as you can. Sketch each arrangement in your notebook. On one side of the page, list arrangements in which the bulb lights. On the other side of the page, list arrangements in which the bulb does *not* light.

B. You should have sketches of at least four different arrangements that light the bulb. How are they similar? How are they different from arrangements in which the bulb fails to light?

C. State what requirements must be met in order for a bulb to light.

An arrangement of a bulb, a battery, and a wire that allows the bulb to light is said to be a *closed electric circuit.* The terms *complete circuit,* or just *circuit* are also used. The word "circuit" was originally used to mean "a circular route or course."

Exercise 1.2

Why is *circuit* an especially suitable name for an arrangement of bulb, battery, and wire in which the bulb is lit?

Experiment 1.3

Examine a flashlight. Make a sketch of the flashlight showing the circuit that exists when the bulb is lit. How many wires are used to make this circuit?

Exercise 1.4

Write an operational definition of an electric circuit. (You may need to refer to the discussion of operational definitions in *Properties of Matter* in Volume I.)

✔ Discuss your definition with a staff member.

Experiment 1.5

Using a bulb, a battery, and two wires, set up an electric circuit in which the bulb is lit. Does it matter which part of the bulb is connected to the end of the battery with the plus sign on it?

Conductors and insulators

Substances can be divided into categories based on their effect on an electric circuit. In

the following experiment we classify some common substances.

Experiment 1.6

Use a battery, a bulb, and two wires to make a circuit in which the bulb lights. Obtain objects made out of the following materials: copper, paper, iron or steel, porcelain, plastic, nichrome wire, glass, aluminum, rubber, and pencil lead. Insert one of these objects into the circuit.

Does the bulb continue to glow brightly or does it dim or go out?

Classify your materials into different categories according to their effect on the bulb. Make a list of the objects in each category.

What do most objects that let the bulb light have in common?

An object that allows the bulb to glow brightly is called a *conductor*. An object that makes the bulb go out is called an *insulator*. Some objects, like pencil lead, fall between the two categories.

Exercise 1.7

Suppose you have a closed box from which two wires protrude. Explain how to use a battery and a bulb to find out whether there is an electrical connection between the two wires inside the box.

Experiment 1.8

Carefully examine a bulb. Use a magnifying glass if possible. You may also find it helpful to look at a broken bulb. Make a careful sketch of the inside of the bulb.

Use your test circuit from Experiment 1.6 to determine whether each part of the bulb is a conductor or an insulator and label it on your sketch. Describe what you believe is the purpose of each part.

Note the two wires coming up from inside the base of the bulb. Use your test circuit to determine where in the base each of these wires originates.

✔ Check your results with a staff member.

Experiment 1.9

A. For convenience, light bulbs are usually placed in sockets. Carefully examine a socket. Identify the conducting and insulating parts, and label these on a careful sketch of the socket. Use the method you invented in Exercise 1.8 to determine which of the conducting parts are connected to one another and show the connections in your sketch.

If there is more than one type of socket available, repeat this experiment for each type. Identify which parts of one type of socket correspond with which parts of other types.

B. Repeat the experiment for a battery holder and switch. Try the switch both open and closed. What do you think is the function of a switch?

C. Using a battery, a holder, a bulb, a socket, and two wires, set up a circuit that lights the bulb. Trace the path of conductors around the circuit. Draw a sketch of the circuit in which you show in detail the conducting path through the socket.

⊘ *Caution:* Do not leave any circuit connected longer than necessary to observe bulb brightness. Leaving a circuit connected for too long can ruin the battery.

Exercise 1.10

Consider the following dispute between two students:

Student 1: *"Here is how a socket works. When the bulb is screwed in, it makes contact with the conductors in the socket as shown in the diagram. The socket thus provides access to the electricity in a safe manner without exposing the user to the electricity."*

Student 2: *"No, that's not how a socket works. My diagram shows how the socket really works. The screw threads on the bulb serve a dual purpose-to hold the bulb in place and to make a connection with the wires. The rivet at the bottom just holds the socket together."*

Do you agree with student 1, student 2, or neither?

✔ Explain your reasoning to a staff member.

Experiment 1.11

A. Connect a battery, a bulb, and a switch so that they form a single closed loop.

Observe the appearance of the bulb as you open and close the switch. Under which circumstances is the circuit not complete? A circuit that is not complete is called an *open circuit.*

B. Connect a switch and a bulb together so that they form a closed loop. With the switch in the open position and still connected to the bulb, connect the ends of the switch to the ends of the battery.

Observe the appearance of the bulb as you close and open the switch quickly. *Do not leave the switch closed!*

🛇 *Caution:* The battery may be ruined if the switch is left closed for a long time in a circuit in which closing the switch makes the bulb go out.

You have just seen that even when a bulb is in a complete circuit, it will not light if there is a wire, or switch, connected across it. In such cases, the wire, or switch, is called a *short* or a *short circuit,* and the bulb is said to be "shorted out."

C. Summarize the conditions under which closing a switch turns a bulb on and the conditions under which closing a switch turns a bulb off.

McDermott & P.E.G., U.Wash./*Physics by Inquiry*

Circuit Diagrams

Circuit diagrams let us represent a circuit on paper by using symbols for the batteries and bulbs instead of drawing pictures. The symbol for a battery is:

The ends of a battery are called *terminals*. The long line represents the positive terminal of the battery (knob end), the short line represents the negative terminal (flat end).

The knob and screw base of a bulb are also called terminals. The symbol for a bulb is:

Note that although the symbol for a battery shows it has two different terminals, the symbol for a bulb shows no difference between the two terminals.

The symbol for a switch is:

The representation used for wires is more complicated than the simple symbols for batteries, bulbs, or switches. As we have seen, contact through a conductor is just the same electrically as direct contact. Direct contact and connection by a wire are therefore represented by the same symbol: a line or a group of connected lines. In each case below, the diagrams show that points *A* and *B* are electrically connected.

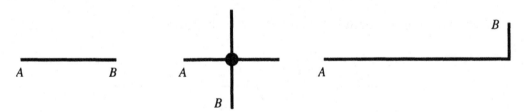

Sometimes it is necessary to draw a circuit in which a wire crosses over another wire, but no electrical connection is made. The symbol used for this is shown at right. In this case the diagram shows that points *A* and *B* are *not* electrically connected.

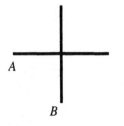

McDermott & P.E.G., U.Wash./*Physics by Inquiry*

Individual parts of a circuit, such as batteries, bulbs, and switches, are often called *circuit elements*. From a circuit diagram, we cannot tell whether two elements of a circuit are far apart and connected by a wire or whether the elements are actually touching. This ambiguity is often confusing because it means that the diagram for a circuit may look quite different from its actual physical appearance. On the other hand, this kind of diagram is particularly useful in analyzing circuits because it focuses on electrical connections rather than the physical arrangement of the circuit.

This last fact is worth emphasizing:

> *Circuit diagrams show electrical connections,*
> *<u>not</u> physical layout.*

We have found that a bulb will light only if there is a complete circuit: a closed loop from one end of the battery, through the bulb, and back to the other end of the battery. A circuit diagram that shows a bulb connected to a battery to form a complete circuit is shown in the accompanying diagram. The wires from the battery are often referred to as positive and negative *leads*.

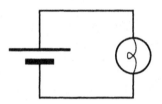

Exercise 1.12

A. In which of the circuits below will the bulb light?

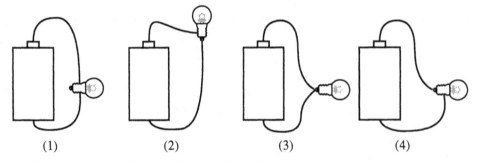

(1)   (2)   (3)   (4)

B. Draw standard circuit diagrams for the four arrangements of battery, bulb, and wires shown in part A.

Which of the circuits in part A have identical circuit diagrams?

Exercise 1.13

A. Draw the circuit diagram for each of the following circuits.

 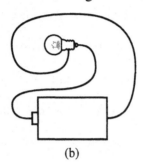

(a)   (b)

B. For which of the following circuits are two wires *required:* only circuit 1, only circuit 2, both circuits, or neither circuit?

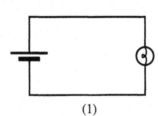   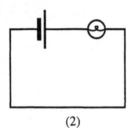

(1)   (2)

C. Can circuit diagram 1 above represent circuit a in part A? Can it represent circuit b?

Can circuit diagram 2 above represent circuit a in part A? Can it represent circuit b?

D. Discuss the ways in which this exercise illustrates that a circuit diagram does not indicate the physical layout of a circuit.

✔ Explain your reasoning to a staff member.

Exercise 1.14

In Experiment 1.1 you found four arrangements of a battery, a bulb, and a single wire that light the bulb. Sketch a circuit diagram for each of the four arrangements.

Which of the four arrangements can be represented by the same circuit diagram? Explain.

Section 2. A model for electric current

In working with electric circuits, you may have noticed some regularities in the way they behave. Perhaps you have begun to form a mental picture, or model, that helps you think about what is happening in a circuit. In this section, we will begin the process of developing a scientific model for an electric circuit.

A scientific model is a set of rules that applies to a particular system that makes it possible to explain and predict the behavior of that system. We would like to build such a model for electric circuits that will enable us to predict the behavior of any circuit of batteries and bulbs. If we connect several bulbs and batteries together in a circuit, we would like to be able to predict which bulbs will light, which will be brightest, dimmest, and so forth.

The first step in building our model will be to incorporate within it features that can account for the behavior we have already observed. As we learn more about electric circuits, we shall add to our model or change it if we need to do so. At any time, we should be ready to discard any part of our model that is in conflict with our observations and that cannot be modified to be consistent with them.

Experiment 2.1

> *Briefly* connect the terminals of a battery with a single wire until the wire feels warm. Does the wire seem to be the same temperature along its entire length or are some sections warmer than others? What might this observation suggest about what is happening in the wire at one place compared to another?
>
> ✔ Check your results with a staff member.

When a wire or a light bulb is connected across a battery, we have evidence that something is happening in the circuit. The wire becomes warm to the touch; the bulb glows. In constructing a model to account for what we observe, it is helpful to think in terms of a flow around a circuit. We can envision the flow in a continuous loop from one

McDermott & P.E.G., U.Wash./*Physics by Inquiry*

terminal of the battery, through the rest of the circuit, back to the other terminal of the battery, through the battery, and back around the circuit. We have found that a light bulb included in this circuit will light. We shall assume that the brightness of the bulb is an indicator of the amount of flow through the bulb.

The *assumptions* that something is flowing through the entire circuit (including the bulb) and that a light bulb can be used as an indicator of the flow are both consistent with our observations. We cannot claim, however, that we have direct evidence for either of these assumptions.

Exercise 2.2

A. In Section 1, you found that a complete circuit was necessary for a bulb to light.

> Does this observation suggest that the flow in an electric circuit is one way (e.g., from battery to bulb) or round trip (e.g., from battery to bulb and back again through the battery)? Explain.

> What does your answer above suggest is a major difference between the flow in an electric circuit and the flow of water in a river?

> Can you tell from your observations thus far the direction of the flow through the circuit?

B. Base your answers to the following questions on the assumptions about the flow in an electric circuit.

> If two identical bulbs are equally bright, what does this indicate about the electric flow through them?

> If one bulb is brighter than another identical bulb, what does this indicate about the flow through the brighter bulb?

Exercise 2.3

Consider the following dispute between two students:

Student 1:   *"When the bulb is lit, there is a flow from the battery to the bulb. There is also an equal flow from the bulb back to the battery."*

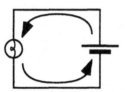

Student 2:   *"The flow is only from the battery to the bulb. We know this is so, because a battery can light a bulb, but a bulb can't do anything without a battery."*

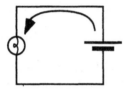

Do you agree with student 1, student 2, or neither?

✔ Explain your reasoning to a staff member.

Since we cannot see anything flowing in an electric circuit, we cannot be sure what kind of flowing process to envision. In keeping with custom, we will use the term *electric current* to refer to the flow. It is important, however, to remember that assigning the name "current" tells us NOTHING about the nature of what flows.

As we build our model step by step, we will draw only on what we can observe in the laboratory and on what we can infer from our observations. A good scientific model is as simple as possible and includes the fewest features necessary for making correct predictions. Thus, we will not include additional features for which we have no direct evidence, such as "electrons." We will find that it is not necessary to identify what flows in an electric circuit in order to predict its behavior.

Thus far we have worked with circuits that contain only one bulb. We will continue developing our model for electric current by investigating what happens in circuits with more than one bulb. Unless stated otherwise, assume all bulbs used in this module are identical. Similarly, assume that all batteries are identical.

Experiment 2.4

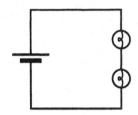

A. Set up a two-bulb circuit with the bulbs connected one after the other as shown.

   Two bulbs connected one after the other are said to be connected in *series*.

   Compare the brightness of each of the bulbs with the brightness of an identical bulb in a single-bulb circuit.

   > Recall the assumptions we have made in developing our model for electric current. How does the current through a bulb in a single-bulb circuit compare with the current through the same bulb when it is connected in series with a second bulb? What does this imply about the current through the *battery?*

B. Compare the brightness of the two bulbs in the two-bulb series circuit with each other. What can you conclude from this observation about the amount of current through each bulb?

   Pay attention to large differences you may observe, rather than minor differences that may occur if two "identical" bulbs are, in fact, not quite identical. (How can you test whether minor differences are due to manufacturing irregularities?)

C. On the basis of your observations and the reasoning you used above, respond to the following questions:

   > Is current "used up" in the first bulb, or is the amount of the flow the same through both bulbs?

   > Do you think the order of the bulbs in this circuit might make a difference? Verify your answer by switching the two bulbs.

Can you tell the direction of the flow through the circuit?

How does the *amount* of the flow through the battery in a single-bulb circuit compare with the flow through the battery in a circuit with two bulbs connected in series?

Exercise 2.5

Consider the following dispute between two students.

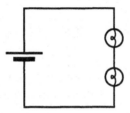

Student 1: *"In this circuit, the flow is from the battery to the first bulb, where some of the current gets used up. Then the rest flows to the second bulb, where all the remaining current gets used up."*

Student 2: *"We know that the current flows back through the battery since we know that we need a complete circuit in order for bulbs to light. If current were used up, there wouldn't need to be a path back to the battery. Furthermore, the bulbs are equally bright so both must have the same amount of current through them."*

Characterize the model of electric current each student is using. Do you agree with student 1, student 2, or neither?

✔ Explain your reasoning for Experiment 2.4 and Exercise 2.5 to a staff member.

In Experiment 2.4, we investigated the behavior of circuits in which two bulbs are connected one after the other in series. We now consider circuits in which the bulbs are connected in a different way.

Experiment 2.6

A. Set up a two-bulb circuit with two identical bulbs so that their terminals are attached together as shown.

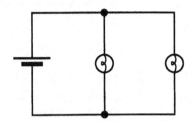

Two bulbs with their terminals attached together in this way are said to be connected in *parallel.*

Compare the brightness of each of the bulbs with the brightness of an identical bulb in a single-bulb circuit.

Recall the assumptions we have made in developing our model for electric current. How does the current through a bulb in a single-bulb circuit compare with the current through the same bulb when it is connected in parallel with a second bulb?

B. Compare the brightness of the two bulbs in the two-bulb parallel circuit with each other. What can you conclude from this observation about the amount of current through each bulb?

Concentrate only on any large differences you may observe, rather than the minor differences that may occur if two "identical" bulbs are, in fact, not quite identical.

C. On the basis of your observations and the reasoning you used above, respond to the following question:

Do you think it is the physical layout of the circuit that makes a difference or the electrical connections? You can investigate this question by comparing what happens:

(1) when the two bulbs are both on the same side of the battery and when they are on different sides.

(2) when each bulb has separate leads to the battery and when the terminals of the bulbs are connected together and then connected to the battery.

D. Describe the flow around the entire circuit for the two-bulb parallel circuit. What do your observations of bulb brightness suggest about the way the current through the battery divides and recombines at the junctions where the circuit splits into the two parallel branches?

McDermott & P.E.G., U.Wash./*Physics by Inquiry*

E. What can you infer about the relative amounts of current through the battery in a single-bulb circuit and in a circuit in which two identical bulbs are connected in parallel across the battery?

F. Does the amount of current through a battery appear to remain constant or to depend on the number of bulbs in a circuit and how they are connected?

Exercise 2.7

Consider the following dispute between two students.

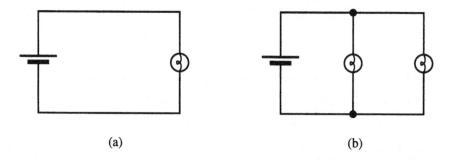

(a)   (b)

Student 1: *"The current through the battery in each circuit is the same. In circuit b the current from the battery is divided between the two bulbs—so each bulb has half the current through it that the bulb in circuit a has through it."*

Student 2: *"We know the current through each of the bulbs in circuit b is the same as through the bulb in circuit a. That's because the bulbs are all about the same brightness —and bulbs that are equally bright have the same current through them. So the flow through the battery in circuit b is more than that through the battery in circuit a."*

Do you agree with student 1 or student 2?

✔ For Experiment 2.6 and Exercise 2.7, explain your reasoning to a staff member.

Section 3. Extending the model for electric current

In Section 2 we began developing a model to help us explain and predict the behavior of electric circuits. According to our model, there is a flow through a complete electric circuit, which we call the electric current. If there is a bulb in the circuit, the brightness of the bulb may be used as an indicator of the amount of the flow through the bulb. The brighter the bulb, the greater the amount of the flow.

Using this model, we found that the current through a battery depends on the number of elements in the circuit and on the arrangement of those elements. In this section we will examine more carefully this dependence of the current on the circuit elements. We also extend our model so that we can predict the behavior of bulbs in circuits with more than two bulbs.

Experiment 3.1

A. Set up a circuit with a battery, a single bulb, and a switch, as shown. Observe the brightness of bulb A.

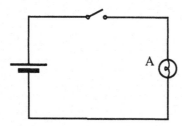

We will consider bulb A to be our *indicator bulb*.

B. Add another bulb of the same type in series as shown. BEFORE you
    close the switch, predict the relative brightness of the bulbs. Do you
    expect the brightness of bulb A to change?

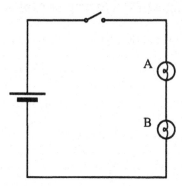

Close the switch.

> How does the brightness of bulb A change, if at all, when another
> bulb is added to the circuit in series with it?

C. Add a third bulb of the same type in series as shown. BEFORE you
    close the switch, predict the relative brightness of the bulbs. Do you
    expect the brightness of bulb A to change?

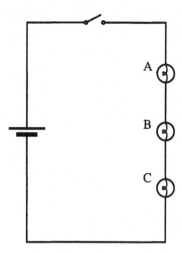

Close the switch.

> How does the brightness of bulb A in this circuit compare with its
> brightness in a single-bulb circuit? How does the brightness of the
> bulb A in this circuit compare with its brightness in a two-bulb
> series circuit?

> Do you think a bulb may have current through it, yet be too dimly
> lit for you to see it?

D. In this experiment you have observed what happens to an indicator bulb (bulb A) as more bulbs are added in series to the circuit. What can you infer about the current through the indicator bulb (and hence through the battery) as each bulb is added in the circuit?

E. We may think of a bulb as presenting an obstacle, or *resistance,* to the current in the circuit. Thinking of the bulb in this way, how would you expect the total obstacle to the flow, or total resistance, to be affected by the addition of more bulbs in series?

F. Use the line of reasoning suggested above to summarize the results of this experiment as a rule that you can include in your model for electric current. The rule should allow you to predict whether the current through the battery increases or decreases as the total resistance in the circuit is increased or decreased.

✔ Explain your reasoning to a staff member.

Exercise 3.2

Suppose you have a closed box in a circuit as shown. You observe a certain brightness in the indicator bulb A. Imagine that someone makes a change in the electrical elements inside the box. You do not see what has been done. However, you observe that the indicator bulb becomes brighter.

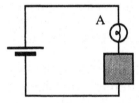

What can you conclude about the current through the battery? Explain how bulb A is serving as an indicator bulb in this circuit.

What can you conclude about the total resistance in the circuit? Has it increased, decreased, or stayed the same? Explain.

Experiment 3.3

A. Set up a circuit consisting of a battery, a switch, and two bulbs, bulb A and bulb B, as shown.

   Bulb A will be our indicator bulb.

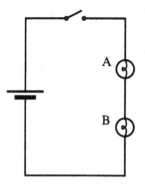

B. Add another bulb of the same type, bulb C, in parallel with bulb B as shown. BEFORE you close the switch, use your model for electric current to predict the relative brightness of the bulbs. Do you expect the brightness of bulb A to change?

   Close the switch.

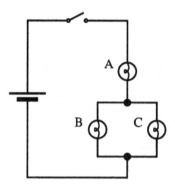

   How does the brightness of bulb A change, if at all, when another bulb is added to the circuit in parallel with bulb B?

C. In this experiment you have observed what happens to an indicator bulb when a bulb is added in parallel with another bulb in the circuit. What can you infer about the current through bulb A (and hence through the battery) as bulb C is added in parallel with bulb B?

   Refer to Exercise 3.2, in which there is a circuit containing a black box. Apply the same line of reasoning to your observations in this experiment. What do you conclude happens to the total resistance of the circuit when bulb C is added in parallel with bulb B?

   Use the results of Exercise 3.2 and this experiment to devise a rule for how the current through the battery changes when a bulb is added in parallel to another bulb.

D. Restate the rule you devised in Experiment 3.1, in which you discussed how adding bulbs in series to a circuit affects the resistance. You should be able to show that the rule you have just devised for bulbs added in parallel to a circuit is consistent with your rule from Experiment 3.1.

✔ Explain your reasoning to a staff member.

McDermott & P.E.G., U.Wash./*Physics by Inquiry*

The results of Experiment 3.3 may seem puzzling. One way to think about the changes in current when the bulbs are connected in parallel is to consider the additional pathways available to the current after it passes through bulb A. Not only is the pathway through the original bulb B still available, but other pathways become available when additional bulbs are connected in parallel with bulb B.

Exercise 3.4

Consider the following dispute between two students:

Student 1: *"Adding bulbs to a circuit increases the total resistance. There is a bigger obstacle to the current so less flows."*

Student 2: *"Adding bulbs to a circuit may increase or decrease the total resistance. It all depends on how you add them. If you add them in parallel, you give the current more pathways, so the total resistance is less."*

Do you agree with student 1 or student 2?

As part of Experiment 2.6, we observed that the brightness of a single bulb connected directly across the battery does not change significantly when a second bulb is connected in parallel to the battery. In the next experiment, we investigate the changes in brightness of bulbs as additional parallel branches are connected across the battery.

Experiment 3.5

A. Set up a single bulb circuit as shown.

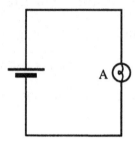

Connect two bulbs in series and add this combination as a branch in parallel with the single bulb as shown.

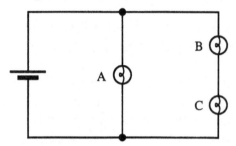

Is there a significant change in brightness of the single bulb when the second branch is connected?

Is there a significant change in brightness of the single bulb or the two bulbs in the second branch when a third branch with three bulbs in series is connected?

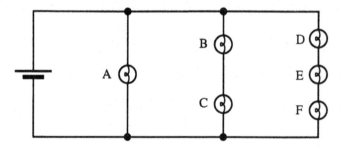

B. Predict what will happen in the following cases. After you have made your predictions, check your answers.

- What will happen to the brightness of bulb E when bulb A is unscrewed? Explain.

- What will happen to the brightness of bulb B and to the brightness of bulb D when bulb E is unscrewed? Explain.

C. We speak of parallel branches as being *independent* of one another when changes in one branch do not significantly affect the other branch. Do parallel branches of bulbs across a battery appear to be dependent or independent of one another?

Experiment 3.6

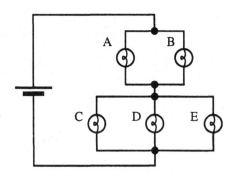

    A. Are the bulbs C, D, and E in the circuit shown connected in series or in parallel?

       Predict the relative brightness of all the bulbs. Explain your reasoning.

       Set up the circuit and test your prediction.

    B. Predict how the brightness of bulbs A and B will *change* when one of the bulbs C, D, or E is unscrewed. Explain.

       Make the changes and test your predictions.

    C. Observe what happens to the brightness of the bulbs C, D, and E when one of them is unscrewed.

       Compare the results of this experiment with the results of Experiment 3.5. Under what conditions are parallel branches independent of one another and under what conditions are parallel branches dependent on one another?

   ✔ Explain your reasoning to a staff member.

Experiment 3.7

    A. Predict the relative brightness of the bulbs in the circuit shown below with the switch open and with the switch closed. Explain your reasoning.

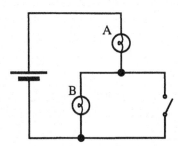

    B. Set up the circuit and check your prediction. If necessary, reconsider your interpretation of what is happening in this circuit.

C. When the switch is closed, bulb B essentially has a wire connected across it.

What does the behavior of bulb B when the switch is closed imply about the way the current divides between bulb B and the wire?

When bulb B is shorted out:

- what happens to the brightness of bulb A?
- how does the current through the battery change?
- how does the total resistance in the circuit change?

D. What does this experiment tell you about the resistance of a wire or switch as compared to that of a bulb? Explain your reasoning.

Considering your observations in Experiment 3.1, would you say that the fact that bulb B appears not to be lit when the switch is closed necessarily means that there is no current through it?

Exercise 3.8

We have inferred from our observations that the current through a battery is not constant but depends on the resistance of the circuit. Formulate a general rule for a qualitative model for electric current in terms of the total resistance in the circuit. Specify how the total resistance depends on the arrangement of the elements. How does adding bulbs in series or parallel affect the current through the battery and through other parts of the circuit?

✔ Check your model for electric current with a staff member.

Exercise 3.9

Consider the circuit shown. The "black boxes" labeled 1 and 2 represent circuit elements. Explain how to use this circuit to compare the resistance of circuit element 1 and circuit element 2.

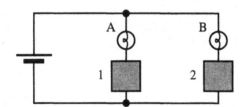

Experiment 3.10

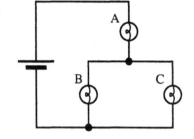

A. Predict the relative brightness of the bulbs in the circuit at right. Explain how you used your model for electric current in making your prediction.

Suppose now that bulb C were unscrewed. Use your model to try to predict what would happen to the brightness of bulb A and bulb B. Explain your reasoning.

Set up the circuit and check your predictions.

B. Consider the following dispute between two students:

Student 1: *"Unscrewing bulb C removes a path for the current. Thus the resistance of the circuit increases and the current through the battery and the remaining bulbs decreases. So bulb A and bulb B would dim."*

Student 2: *"I agree that bulb A will dim, but I disagree about bulb B. Before you unscrew bulb C, only part of the current through bulb A goes through bulb B. Afterward, all of the current through bulb A goes through bulb B. So bulb B should get brighter."*

Would you say that either student has given a complete answer? Explain.

✔ Discuss your results with a staff member.

The preceding experiment illustrates a situation for which a simple application of our present model for electric current is not sufficient to allow us to predict bulb brightness. In Part C of this module, we will extend the model so that we can account for the behavior of bulbs in more complicated circuits.

     McDermott & P.E.G., U.Wash./*Physics by Inquiry*

Section 4. Series and parallel networks

The terms *series* and *parallel* refer to the ways in which bulbs are connected in a circuit. However, there are many circuits in which bulbs are not connected in simple series or in simple parallel combinations. Consider, for example, the circuits below, which we examined in Section 3.

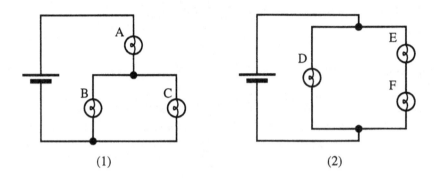

(1)   (2)

To make predictions for these and other more complicated circuits, we need to refine our definitions for series and parallel connections. The following definitions are more precise and should be used in deciding whether two bulbs in a circuit are connected in series or in parallel.

Series connection: Two bulbs are in series if they are connected so the same current that passes through one bulb must also pass through the other.

Parallel connection: Two bulbs are in parallel if their terminals are connected together so that *at each junction* one terminal of one bulb is directly connected to one terminal of the other.

Using these definitions, we see that in circuit 1 above, bulbs B and C are in parallel with each other. Bulb B is not in series with bulb A; instead, the entire unit consisting of bulbs B and C is in series with bulb A. In circuit 2 above, bulbs E and F are in series with each other. Bulb E is not in parallel with bulb D; instead, the entire unit consisting of bulbs E and F is in parallel with bulb D.

Exercise 4.1

Explain why the following two statements are direct consequences of the preceding definitions.

A. When two bulbs are connected in series, they have a single common junction and together, as a unit, constitute the only continuous path through that junction.

B. When two bulbs are connected in parallel, current that passes through one bulb does not pass through the other.

Both of these statements may be useful in helping identify a series or parallel connection, but the real criterion is the definition.

Exercise 4.2

Apply the definitions for series and parallel connections to the following circuits to determine all the series and parallel connections of bulbs. Which of the circuits below are "equivalent," that is, which of the circuits shown can be used to represent the same physical circuits?

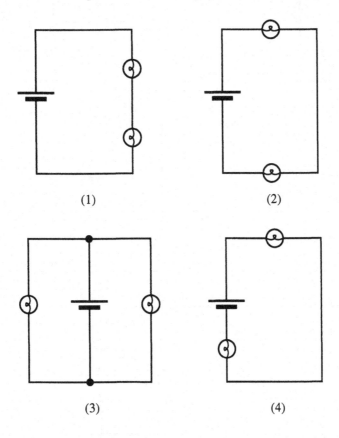

(1)   (2)

(3)   (4)

It is often useful to think of a circuit that consists of more than one series or parallel connection of elements as a combination of *networks.* An electrical network is an assembly of connected elements. In the examples at the beginning of this section, bulbs

B and C in circuit 1 form a parallel network; bulbs E and F in circuit 2 form a series

network. The definitions for series and parallel connections of bulbs can be extended to

include networks of elements.

Exercise 4.3

In the following circuits, determine all the series and parallel connections of bulbs and networks of bulbs. (You may find it helpful to redraw some of the circuits.) Are there any bulbs not connected in series or in parallel with another bulb or network of bulbs?

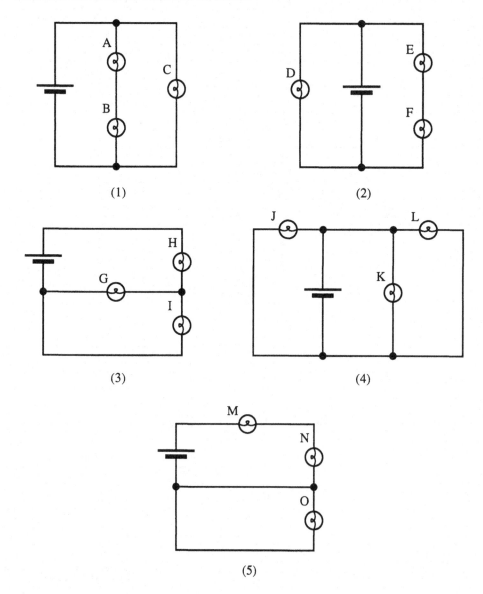

✔ Check your results with a staff member.

McDermott & P.E.G., U.Wash./*Physics by Inquiry*

Exercise 4.4

Suppose you have three boxes, labeled A, B, and C. Each box has two terminals. The arrangement of bulbs inside each box is shown below.

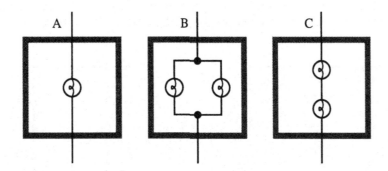

A. For each of the following circuits, draw a standard circuit diagram showing all the bulbs in the circuit. List the series and parallel combinations for each circuit.

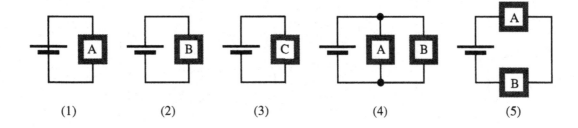

B. Rank each of the circuits in part A according to the current through the battery.

✔ Explain your reasoning to a staff member.

Experiment 4.5

A. With the switches in all possible positions, tell whether the bulbs are in series, in parallel, or neither in each of the following circuits. There are four possibilities for each diagram. In each case, predict the brightness of the bulbs relative to the other bulbs in the circuit. If there are any arrangements that short the battery, state that explicitly.

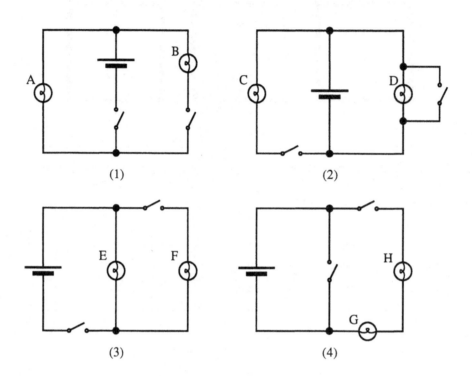

       (1)   (2)

       (3)   (4)

B. Set up the circuits and check your predictions except in the case(s) that would result in a short circuit across the battery. (See Experiment 1.11 for the definition of a short circuit.)

C. Two of the above circuits work in exactly the same way. Which two? Explain how they are alike.

In the next series of experiments, we explore what happens in circuits with more than three bulbs. We can test our qualitative model for electric current by making predictions for the brightness of bulbs in various combinations.

Experiment 4.6

Two students are predicting the brightness of identical bulbs in the circuit below.

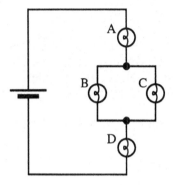

Student 1: *"All the current is through A. Then it divides between B and C so they will be equally dim, lots dimmer than A. Then the current comes together again and it all goes through D. Bulb D will be the same brightness as bulb A."*

Student 2: *"I think D will be a lot dimmer than A; in fact, maybe it won't light at all. There won't be much current left after it passes through A and B and C. Maybe D will be bright and A will be dim; it depends on the direction of the flow through the circuit. This would be a good test to find the direction of the current."*

Do you agree with student 1, student 2, or neither? Explain your reasoning.

Set up this circuit and observe the relative brightness of the bulbs.

Experiment 4.7

Set up the circuit shown at right.

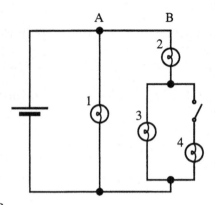

A. Observe the change in brightness of bulb 2
    when the switch is opened and closed.
    Now observe bulb 1 as the switch is opened
    and closed.

    Which of these two bulbs has a clearly
    observable change in brightness? Which
    has no noticeable change or a change so
    small that it can be considered negligible?

B. Notice that there are two branches of the circuit, labeled A and B, that
    are connected in parallel directly across the battery. Several alterations
    to each branch are listed below. In considering the effect of each
    alteration, begin with the circuit in its original state.

    (1) Predict the effect on branch A of each of the following alterations
         to branch B:

         •   unscrewing bulb 2

         •   shorting out bulb 3

         Explain your reasoning.

    (2) Predict the effect on branch B of each of the following alterations
         to branch A:

         •   unscrewing bulb 1

         •   adding a bulb in series with bulb 1

         Explain your reasoning.

C. Make each alteration and record your observations. How would you account
    for any differences between your predictions and observations?

D. In this experiment two parallel branches are connected directly across
    the battery. On the basis of your observations would you say that
    changes in one branch affect the other or that the two branches appear
    to be independent of each other?

✔ Check your results with a staff member.

Experiment 4.8

A. Without setting up the circuit at right, use your model for electric current to predict the following:

• How will the brightness of bulb A compare to the brightness of bulb B?

• How will the brightness of bulb B compare to the brightness of bulb D?

• How will the brightness of bulb A compare to the brightness of bulb C?

Explain your reasoning in each case.

B. Set up the circuit and observe the actual brightness of the bulbs. Explain any differences between your predictions and your observations.

✔ Check your reasoning with a staff member.

Exercise 4.9

A. Which of the circuit diagrams below can be used to represent the same physical circuits; that is, which circuits have the same electrical connections? To make this decision you may find it helpful to redraw some of the circuits.

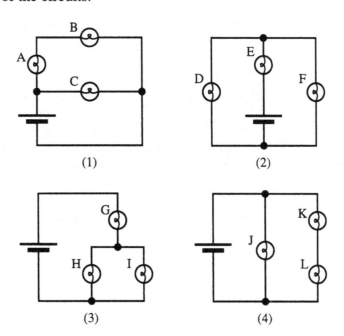

(1)   (2)

(3)   (4)

B. How many different circuits are represented by the diagrams above? In each case identify the series and parallel connections of bulbs and networks of bulbs.

The preceding exercise illustrates that a circuit may be drawn in a variety of ways. Often, by redrawing a complicated circuit, we can more easily identify the series and parallel connections.

Exercise 4.10

Below are two circuit diagrams drawn in unusual ways.

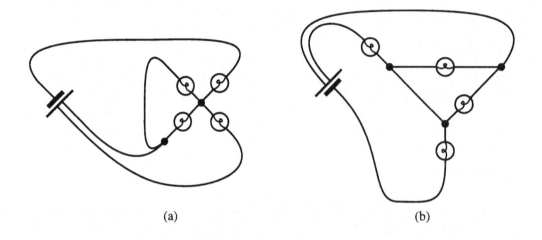

(a)   (b)

Pick the standard diagrams below that represent the circuits above.

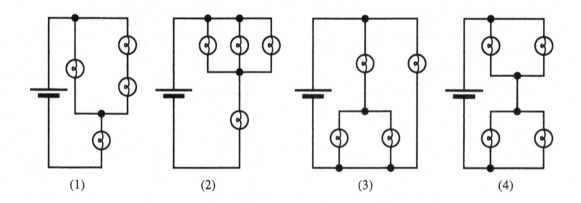

(1)   (2)   (3)   (4)

Exercise 4.11

Below is a circuit diagram drawn in an unusual way.

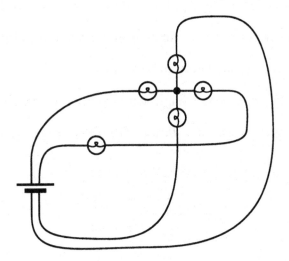

Pick the standard circuit diagram below that represents the circuit above.

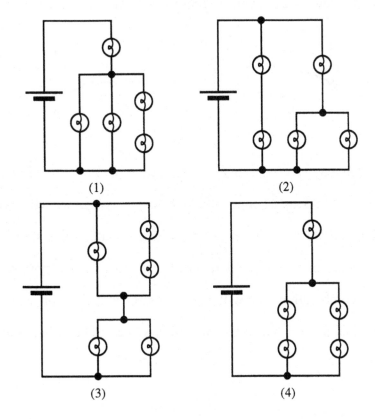

(1)   (2)

(3)   (4)

✔ Explain your reasoning to a staff member.

Switches

Switches allow us to control the current in an electric circuit. The type of switch we have been using until now is called a *single-pole–single-throw switch,* or an SPST switch. An SPST switch can be either in a closed or in an open position in a circuit and thus either make or break a single connection between two wires.

Another type of switch connects a single wire to either of two other wires. This type of switch is called a *single-pole–double-throw (SPDT) switch.* The symbol for an SPDT switch is:

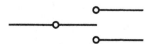

An SPDT switch has three terminals. These are labeled *A, B,* and *C* in the two diagrams below. The switch can connect terminal *C* to either terminal *A* or terminal *B.*

An SPDT switch may be used to make a connection in either one of two networks while simultaneously breaking a connection in the other. When we work with circuit diagrams containing SPDT switches, the symbol for the switch often shows terminal *C* as unconnected to either *A* or *B.* For the purpose of predicting bulb brightness and finding series and parallel combinations, we will *always* assume that *C* is connected to either terminal *A* or terminal *B.* In other words, we will assume SPDT switches are never open.

Experiment 4.12

A. Predict whether each bulb (X or Y) will light in the following circuit for each of the switch positions listed below. In each case, state whether the bulbs are in series, in parallel, or neither.

- Switch 1 at position *A*, switch 2 at position *C*.

- Switch 1 at position *A*, switch 2 at position *D*.

- Switch 1 at position *B*, switch 2 at position *C*.

- Switch 1 at position *B*, switch 2 at position *D*.

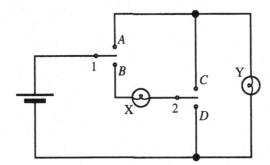

✔ Set up this circuit and demonstrate your predictions to a staff member.

Experiment 4.13

Sometimes it is desirable to be able to control a single bulb from two locations (e.g., as a stairway light might be controlled by switches at the top and bottom of the stairs). Design a circuit using two SPDT switches, a bulb, and a battery so that either of the switches can turn the bulb on or off regardless of the position of the other switch. Be careful that the circuit you design does not short the battery for one of the switch positions.

✔ Set up the circuit and demonstrate it for a staff member.

Exercise 4.14

Ask a staff member to show you the circuit for the exercise.

A. Draw a standard circuit diagram for this circuit.

B. For each of the possible switch positions, identify which bulbs are connected in series and which are connected in parallel.

C. For each of the possible switch positions, predict the brightness of the bulbs.

Test your predictions.

Part B: Measurements of current and resistance

In Part A of this module, we used the brightness of a light bulb as an indicator for current. We assumed that the brighter the bulb, the greater the current through it. This assumption was sufficient for predicting the brightness of identical bulbs in various circuits. As long as we are concerned only with relative brightness, bulbs provide a convenient means for studying the behavior of electric circuits.

In Part B, we begin a more quantitative analysis of electric circuits. Ammeters are used to measure the magnitude of the current through the elements of a circuit. We investigate how the resistance of a network depends on the way in which its elements are connected to one another.

Section 5. Kirchhoff's first rule

In a closed circuit, there is a continuous path from one terminal of the battery, around the circuit to the other terminal, and then back through the battery to the first terminal. On the basis of our observations and assumptions, we have determined that current is not "used up" in the circuit. Another way of expressing this idea is to say that current is "conserved." The *conservation of current* is a fundamental principle that holds for all closed circuits. In this section, we examine this idea and express it in quantitative terms.

We cannot easily quantify current by using a light bulb. We will use *linear resistors;* to investigate the conservation of current. (Linear resistors have a resistance essentially independent of the current through them. As we will see later in this module, bulbs do not have this property.)

We will be using linear resistors made of *nichrome* wire. The symbol for a resistor is:

To measure current, we will use an instrument called an *ammeter,* which allows current to pass through it without altering the resistance of the circuit very much. When an ammeter is connected in series in a circuit, it measures the current through the circuit in terms of a unit called an *ampere* (A). Often it is convenient to use a smaller unit, such as a *milliampere* (mA), which is one-thousandth of an ampere. The symbol we will use for current is *i.*

We have found that it is impossible from our observations to tell the direction of the current through the battery. We will follow the widely used convention of *assuming* that the flow is from the positive terminal of the battery through the circuit to the negative terminal, and from the negative to the positive terminal within the battery. The actual direction of flow is unimportant, but it is very important to be consistent. Ammeters and other electrical instruments must be connected in a circuit in the same sense as the battery, otherwise they may be damaged.

The symbol for an ammeter is:

One terminal of the ammeter is marked positive; the other negative. Some ammeters have more than one positive or negative terminal.

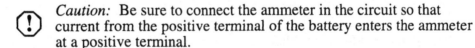

Caution: Be sure to connect the ammeter in the circuit so that current from the positive terminal of the battery enters the ammeter at a positive terminal.

If the ammeter has more than one positive or negative terminal, it is a good practice to complete the circuit by using the terminal marked with the largest scale first and working down to the most sensitive scale for which the needle will remain on the scale. (This procedure is a precaution to protect the ammeter. If the ammeter has been connected backwards, use of the least sensitive scale will minimize the damage.)

Experiment 5.1

A. Connect a circuit containing a battery, bulb, and ammeter as shown. Record the reading on the ammeter.

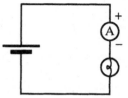

B. Predict the reading on the ammeter in the second circuit shown at right. Explain your reasoning.

Connect the circuit and check your prediction.

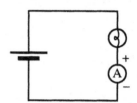

How do the readings on the ammeter in the two circuits compare? Is this observation consistent with your model for current? Explain.

Experiment 5.2

A. Obtain three 30 cm lengths of nichrome wire. Connect an ammeter, one piece of nichrome wire, a battery, and a switch in series, as shown. Record the reading on the ammeter.

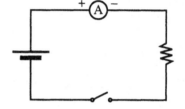

Repeat the procedure with two pieces of nichrome wire connected in series, and then with three. Record your measurements.

What happens to the current when the length of nichrome wire is doubled and tripled? Is your observation consistent with the idea that the nichrome wire has resistance?

B. Obtain a 90 cm length of nichrome wire. Connect an ammeter, the nichrome wire, the battery, and the switch in series as shown. The arrow in the diagram represents a movable contact that is made by sliding the lead from the battery along the nichrome wire.

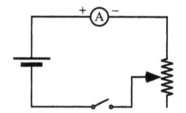

A resistor with a movable lead is often called a *variable resistor.*

Place the movable lead so that there is a 30 cm length of wire in the circuit. Record the reading on the ammeter. Repeat this procedure with 60 and 90 cm lengths of nichrome wire in the circuit.

C. Compare the ammeter readings from part A with the corresponding readings from part B.

How does using individual pieces of nichrome wire connected in series compare to using a single piece of nichrome wire with a movable lead?

Does the portion of the nichrome wire that extends beyond the lead from the battery have any effect on the circuit?

D. Write an algebraic equation that relates the current in the circuit with 30 cm of nichrome wire, i_1, to the current in the circuit with 60 cm of wire, i_2. Interpret this equation by stating whether i_1 is larger or smaller than i_2 and by what factor.

Write an algebraic equation that relates the current in the circuit with 30 cm of wire to the current in the circuit with 90 cm of wire, i_3. Interpret this equation by stating whether i_1 is larger or smaller than i_3 and by what factor.

Write an algebraic equation that compares the current in the circuit with 60 cm of wire to the current in the circuit with 90 cm of wire. Interpret this equation by stating whether i_2 is larger or smaller than i_3 and by what factor.

Reverse the comparison between the 60 cm and 90 cm cases by comparing the current in the circuit with 90 cm of wire to the current in the circuit with 60 cm of wire. Interpret this equation by stating whether i_3 is larger or smaller than i_2 and by what factor.

✔ Check your results with a staff member.

Experiment 5.3

Connect an ammeter, a battery, and a switch in series with two 15 cm lengths of nichrome wire that form a parallel network as shown. Record the measurement.

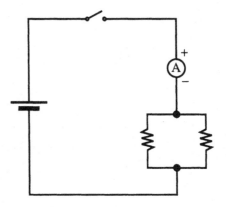

Disconnect the ammeter and reconnect it to the other side of the parallel network. Record the measurement.

What do you conclude about the magnitude of the current on both sides of the parallel network?

Connect the ammeter to each of the parallel branches of the network. Check your connections with a staff member, then record each measurement.

What do you conclude about the magnitude of the current in the branches compared to the magnitude on either side of the parallel network?

✔ Check your results with a staff member.

    McDermott & P.E.G., U.Wash./*Physics by Inquiry*

As the preceding experiment illustrates, the magnitude of the current may change at the locations where circuit elements are connected together. These points, such as points A and B in the following circuit, are called *nodes*. Whenever circuit elements are connected, they form a node.

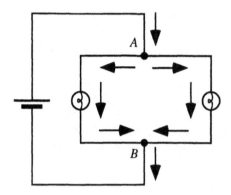

Point A is called a three-way node because there are three ways to get to or away from point A. Point B is also a three-way node. Because conductors can be represented on circuit diagrams by various sets of lines, a node may be represented in many different ways. The following, for example, are all equivalent representations of a three-way node.

Identifying the nodes can be helpful in drawing the circuit diagram for a circuit that is already set up. A good procedure for making a diagram for an existing circuit is the following: Begin tracing the circuit from the battery and proceed until the first node is encountered. Then draw in the circuit element along each path out of the node. Choose one branch and follow it to the next node. Each time a node is encountered, stop and draw in the circuit elements connected to the node.

Exercise 5.4

How many nodes are there in the following circuit?

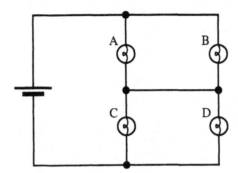

Predict the relative brightness of the bulbs in the circuit, and tell whether the bulbs are arranged in series or in parallel.

Exercise 5.5

In Experiment 5.3, you used an ammeter to find the current at several locations in the circuit shown at right.

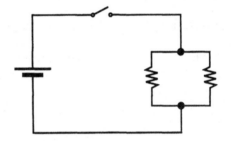

A. Identify the nodes in this circuit.

B. Find the current into each node and the current out of each node.

How does the current into a node compare to the current out of the node? How might you express this result mathematically?

We are now ready to use the idea of current conservation from our model. One way of expressing this idea is in terms of the relation between the total current into a node and the total current out of the node. For example, consider the following node. (The symbol used to represent current algebraically is i.)

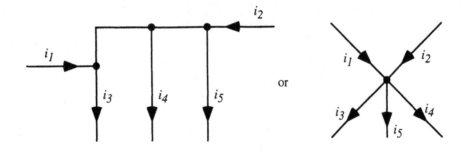

In the previous diagram, a five-way node is shown. The direction of two of the currents is towards the node, and the direction of three of the currents is away from the node. Conservation of current requires:

$$i_1 + i_2 = i_3 + i_4 + i_5$$

The conservation of current at a circuit node is called *Kirchhoff's first rule,* and is stated below.

The total current out of a node is equal

to the total current into the node.

This statement can also be expressed in terms of the algebraic sum of the currents. From the preceding equation that expresses the conservation of current at the five-way node in the diagrams, we obtain:

$$i_1 + i_2 - i_3 - i_4 - i_5 = 0$$

This form of the equation leads to the following alternate statement for Kirchhoff's first rule:

The algebraic sum of the currents

at a node is zero.

Expressing Kirchhoff's first rule as an algebraic sum is equivalent to assigning a positive sign to current into a node and a negative sign to current out of a node (or vice versa). In terms of amperes, Kirchhoff's first rule can be stated as: *The total number of amperes flowing into a node is equal to the total number of amperes flowing out of the node.*

Exercise 5.6

The circuit at right consists of a battery and four identical bulbs. The arrows indicate the assumed direction of the current through certain elements.

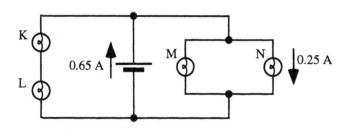

Find the current through each bulb.

Exercise 5.7

The circuit shown at right contains several different electrical elements. (The different shapes represent different kinds of elements.)

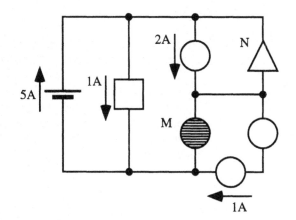

Find the current through elements M and N.

Exercise 5.8

A total current i divides into two branches. One branch carries 15% more current than the other.

Write algebraic expressions in terms of i for each of the currents.

Exercise 5.9

In the arrangement of elements shown at right, 6 A and 3 A flow in, and 5 A and 4 A flow out. Suppose that element K has 1.7 times as much current through it as element N.

Find the current through elements K, L, M, and N. (*Hint:* Use variables to represent these currents and apply Kirchhoff's first rule at various nodes to yield equations relating the four currents. Then solve for the currents.)

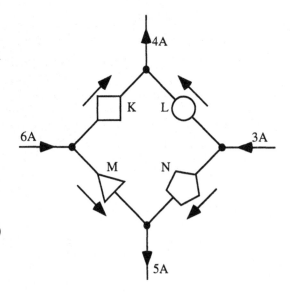

Section 6. Equivalent resistance

In our investigations of electric circuits we have found it worthwhile to think of bulbs and

other elements as presenting an obstacle, or resistance, to the current in a circuit. We will

find the idea of resistance more useful if we can determine a way to measure it. In this

section we develop an operational definition for resistance.

Experiment 6.1

A. In the circuit shown at right, the resistor C
represents a short piece of nichrome wire.

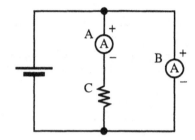

Which ammeter will have a greater
current through it? Explain your
reasoning. (*Note:* Do not set up this
circuit; one of the ammeters could be
damaged.)

B. In the circuit shown at right, the variable
resistor D represents a piece of nichrome wire
whose length we can vary.

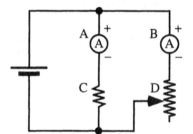

What would you do to the variable resistor D
to make the current through ammeter A:

(1) greater than the current through
ammeter B?

(2) equal to the current through ammeter B?

(3) less than the current through ammeter B?

Set up the circuit and check your answer.

C. Describe a way that you could use the circuit in part B to measure the
size of an unknown obstacle. What units would you use to express the
size of the obstacle?

Exercise 6.2

Experiment 6.1 suggests a way that we can measure resistance. Write
down a clear and unambiguous operational definition for resistance.

✔ Check your operational definition with a staff member.

The operational definition for resistance in Exercise 6.2 can be used to determine not only the resistance of individual elements, but also the resistance of networks of elements. In the experiments that follow, we examine how resistances combine for elements connected in series and in parallel.

Experiment 6.3

A. Apply your operational definition for resistance to measure the resistance of the following series networks of resistors.

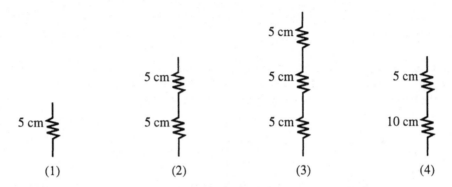

Based on your observations, try to come up with a mathematical rule for how resistances in series combine. (*Note:* It is difficult to obtain exact answers with this equipment. Try to look for general patterns.)

Is your rule consistent with your qualitative rule from Section 3 for how resistances in series combine? Explain.

B. Apply your operational definition for resistance to measure the resistance of the following parallel networks of resistors.

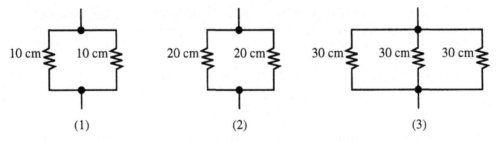

Based on your observations above, try to come up with a rule for how resistances in parallel combine. (*Note:* It is difficult to obtain exact answers with this equipment. Try to look for general patterns.)

Is your rule consistent with your qualitative rule from Section 3 for how resistances in parallel combine? Explain.

✔ Check your results with a staff member.

          McDermott & P.E.G., U.Wash./*Physics by Inquiry*

In the preceding experiments you used an ammeter to find the length of nichrome

wire that presented the same obstacle to the current as a given network of wires. The

single wire and the network of wires are equivalent in that one can be switched for the

other without affecting the current in the rest of circuit. The resistance of each network

that you measured in Experiment 6.3 is called the *equivalent resistance* of that network.

Note that, in general, the equivalent resistance of a network is different from the

resistance of the individual elements that make up the network.

Exercise 6.4

A. Use the rules you developed in Experiment 6.3 to calculate the equivalent resistance of each of the following arrangements of nichrome wire. Do not make any measurements.

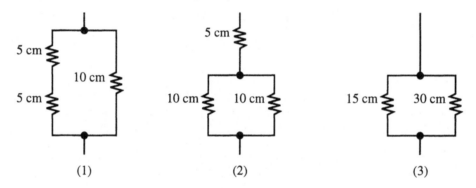

        (1)   (2)   (3)

(*Hint for network 3:* how can you represent the 15 cm length of wire as a combination of 30 cm pieces of nichrome wire?)

B. Suppose you had several 5 cm lengths of nichrome wire but no others. Without cutting any of the wires, how could you make networks with the following resistances:

   (1) a resistance equivalent to 10 cm of nichrome wire.

   (2) a resistance equivalent to 7.5 cm of nichrome wire.

   (3) a resistance equivalent to 1 cm of nichrome wire. (*Note:* Simply clipping the leads 1 cm apart along a single 5 cm length of wire does not count!)

Experiment 6.5

A. Set up the circuit shown at right and find the length of nichrome wire that results in the same current through both branches of the circuit.

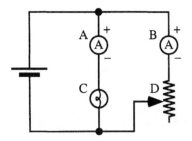

Replace bulb C with a network of two bulbs in series. Find the length of nichrome wire that results in the same current through both branches.

Replace the two bulbs in series by three bulbs in series and repeat the procedure described above.

B. In part A of Experiment 6.3, you performed an experiment similar to that in part A above, but used identical pieces of nichrome wire instead of identical bulbs. Compare the results of the two experiments. How do your observations in the two experiments differ?

C. Explain how the observations in this experiment are consistent with a view of bulbs as having a resistance that changes when the current through the bulb changes.

Does the resistance of a bulb increase or decrease when the current through it increases? Explain your reasoning. Base your answer on your results from this experiment and the rule for adding resistances that you developed in the previous experiment.

✔ Check your reasoning with a staff member.

McDermott & P.E.G., U.Wash./*Physics by Inquiry*

Part C: Measurement of voltage

In Part C, we examine the effect of adding batteries to a circuit. We develop a general method for the analysis of circuits that contain several batteries, bulbs, and resistors.

Section 7. Multiple batteries

According to our model, when the terminals of a battery are connected in a complete circuit, there is a current through the circuit. For a given battery, the magnitude of the current depends on the total resistance of the circuit. In the following experiments, we examine what happens when additional batteries are added to a circuit.

Experiment 7.1

A. Set up a single-bulb circuit, then add a second bulb in series.

How does the brightness of each of the bulbs in the two-bulb circuit compare with the brightness of the single bulb?

B. Connect a second identical battery as shown.

Two batteries, connected so that the negative terminal of one is attached to the positive terminal of the other, are said to be in *series*.

What happens to the brightness of the two bulbs when the second battery is added?

How does the brightness of the bulbs in this circuit compare with the brightness of the single bulb in the one-battery circuit?

C. Repeat the above procedure with a third identical bulb and a third identical battery.

How does the brightness of the bulbs in the three-battery circuit compare with the brightness of the single bulb in the one-battery circuit?

How do the currents in the one-battery–one-bulb, two-battery–two-bulb and three-battery–three-bulb circuits compare? Explain.

As the resistance in an electric circuit increases, we find that we must increase the number of batteries connected in series for the current to remain the same. This observation leads us to think of the battery as the agent that "pushes" or "drives" the current through the circuit. In the experiment below, we examine the relationship between the number of batteries in series and the magnitude of the current.

Experiment 7.2

Connect a battery in a complete circuit with a single bulb. Observe the brightness of the bulb.

Connect a second battery in series with the first. Observe the brightness of the bulb.

Connect a third battery in series with the first two. Observe the brightness of the bulb. Do not leave the bulb in the circuit any longer than necessary to make your observation.

What happens to the brightness of the bulb when additional batteries are added to the circuit?

How does increasing the number of batteries in the circuit affect the current?

We can obtain a measure of the strength of the battery by using an instrument called a *voltmeter*. When the terminals of a voltmeter (usually marked positive and negative) are connected to the

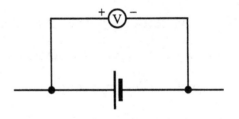

corresponding terminals of a battery, a needle on a scale deflects from its zero position. The voltmeter reading is an indicator of the strength of the battery, that is, the ability of the battery to drive current through a circuit. The number obtained from this measurement is expressed in *volts* (V) and is commonly referred to as the *voltage* of the battery, or more precisely, as the "voltage *across* the battery," or "the *magnitude* of the voltage across the battery."

Experiment 7.3

A. Measure the voltage across a battery by connecting the positive terminal of the voltmeter to the positive terminal of the battery and the negative terminal of the voltmeter to the negative terminal of the battery.

   Compare your voltage measurement with the voltage rating indicated on the battery. Are these numbers equal (or almost equal), or do they differ significantly?

B. How would you connect the voltmeter so it will measure the voltage across two batteries connected in series? Check with a staff member before setting up your circuit.

   How does the number obtained from the measurement compare with the voltage indicated on the batteries?

C. Measure the voltage across three batteries connected in series.

   How does the number obtained from the measurement compare with the voltage indicated on the batteries?

D. How do the numbers you obtain for the voltages across the various combinations of batteries compare with the ability of the combinations of batteries to drive current through a circuit? Support your answer with your observations from Experiments 7.1 and 7.2.

A voltmeter will indicate a voltage not only when it is connected across a battery, but also when it is connected across any circuit element that has current through it. A voltmeter is designed so that it can measure the voltage across an element without significantly affecting the voltages in the circuit.

To measure the voltage across a circuit element, the voltmeter is connected in parallel with it. In the diagram below, the reading on the voltmeter indicates the voltage across the resistor.

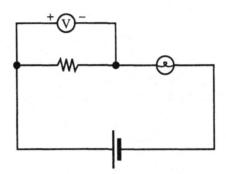

In using a voltmeter to measure voltages, we must take certain precautions. A common type of damage results from connecting the leads "backwards," so that the needle is forced downward past zero. Good meters are able to withstand a certain amount of this kind of abuse, but they will eventually be damaged. The positive terminal of the voltmeter should be connected to the terminal of the circuit element that is electrically closer to the positive terminal of the battery. When it is not clear which way to connect the voltmeter leads, connect just one lead, then tap the second lead in place to make fleeting contact while you are watching the meter. If the needle jumps the wrong way, reverse the leads.

A voltmeter can also be damaged by trying to measure a voltage greater than the meter can indicate. If the maximum voltage on the meter scale is 5 volts and we attempt to measure a voltage of 100 volts, the meter is likely to be irreparably damaged. Just like in using the ammeter, it is a good habit always to use the highest voltage scale on the voltmeter first, then, if necessary, to switch to lower voltage scales, step by step.

Experiment 7.4

Obtain a 15 cm length of nichrome wire, a battery, a bulb, and some connecting wire.

A. Before setting up a circuit, measure the voltage across the battery, bulb, and each piece of wire. What is the voltage across each of them when not in a complete circuit?

B. Consider now a circuit with the bulb, the nichrome wire, and the length of connecting wire all attached in series across the battery. Draw four separate diagrams to show how you would connect a voltmeter to measure:

  • the voltage across the battery.

  • the voltage across the nichrome wire.

  • the voltage across the connecting wire.

  • the voltage across the bulb.

  (!) Check with a staff member to be sure that you are planning a procedure that will not damage the meter.

C. Set up the circuit, and make the measurements.

  Does the voltage across some of the elements remain the same (or almost the same) whether or not they are in a complete circuit?

  Does the voltage across some of the elements change when they are placed in a complete circuit? If so, does the voltage increase or decrease?

D. Place the four elements above into categories based on their behavior both in and out of electric circuits.

  ✔ Discuss the similarities and differences in behavior of these four elements with a staff member.

Experiment 7.5

A. Imagine a circuit with a single battery and bulb. How would you expect the measurements for the voltage across the battery and bulb to compare?

Set up the circuit, make the measurements, and check your predictions.

B. In this part of the experiment, you will place the voltmeter leads at different locations along the connecting wires and investigate the effect on the voltmeter reading.

(1) Reconnect the voltmeter so it measures the voltage across the bulb. Note the voltmeter reading. Now, move the negative lead of the voltmeter so it is attached to the negative terminal of the battery.

How does this affect the reading on the voltmeter? Would this be a valid way to measure the voltage across the bulb?

(2) Return the negative lead of the voltmeter to its original position, and move the positive lead of the voltmeter to the positive terminal of the battery.

How does this change affect the reading on the voltmeter? Would this be a valid way to measure the voltage across the bulb?

(3) Now, move the negative lead of the voltmeter to the negative terminal of the battery.

How does this change affect the reading on the voltmeter? Would you say this was a valid way to measure the voltage across the bulb? Would you say this was a valid way to measure the voltage across the battery?

C. In light of your results from this experiment, justify the following statement:

To measure the voltage across an element, the voltmeter leads may be placed anywhere along the connecting wire attached to the appropriate terminals of the element.

✔ Explain your reasoning to a staff member.

   McDermott & P.E.G., U.Wash./*Physics by Inquiry*

Experiment 7.6

A. Set up a circuit with one 30 cm length of nichrome wire connected across a battery.

> Predict how the voltage across the battery compares to the voltage across the nichrome wire. Make the measurements and check your prediction.

B. Add a second identical piece of nichrome wire to the circuit. Connect the second piece in series with the first piece of wire.

> Predict how the voltage across the first piece of nichrome wire compares to the voltage across the other piece of nichrome wire.

> Predict how the voltage across the battery compares to the voltage across the combination of the two nichrome wires.

Set up the circuit and check your predictions.

C. Add a third identical piece of nichrome wire to the circuit in series with the other lengths of nichrome wire.

> Predict how the voltages across the three pieces of nichrome wire compare.

> Predict how the voltage across the battery compares to the voltage across the combination of three pieces of nichrome wire.

Use a voltmeter to check your predictions.

D. Find an algebraic equation relating the voltage across the battery and the voltages across the other elements in the circuit.

In Experiment 7.1, we found that as the total resistance in a circuit increases, if we want to keep the current constant, we must increase the voltage by adding batteries. In Experiment 7.2 we found that as we add more batteries to a circuit, the current through the circuit increases. On the basis of these observations, we extended our model for electric current to include the idea that the battery causes the flow in a circuit.

In Experiment 7.4, we observed that circuit elements other than a battery do *not* have a voltage across them *unless* they are connected as part of a complete circuit. For a single element connected across a battery, we found in Experiment 7.5 that the voltage across the element equals the voltage across the battery. In Experiment 7.6, we found that for several elements connected in series across the battery, the sum of the voltages across the elements equals the voltage across the battery.

We now will examine in more detail the relationship between the voltage across an element in a circuit and the current through that element.

Experiment 7.7

Set up a circuit with a battery, a bulb, an open switch, and an ammeter connected in series. Then connect a voltmeter in parallel with the bulb as shown. Before closing the switch, be sure to take the necessary precautions to protect the meters.

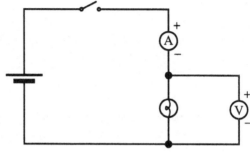

We shall use the ammeter to measure the current through the bulb and the voltmeter to measure the voltage across the bulb.

A. Close the switch. Read the ammeter and voltmeter. Now, remove the ammeter from the circuit. Is the reading on the voltmeter significantly different? Does the brightness of the bulb change significantly?

   What do you infer about the resistance of the ammeter? Is it relatively large or small?

B. Replace the ammeter in the circuit, and remove the voltmeter. Is the reading on the ammeter significantly different? Does the brightness of the bulb change significantly?

   What do you infer about the resistance of the voltmeter? Is it relatively large or small?

C. Does the presence of the voltmeter or the ammeter, when properly connected to a circuit, significantly affect the conditions in the circuit?

✔ Explain your reasoning to a staff member.

McDermott & P.E.G., U.Wash./*Physics by Inquiry*

Experiment 7.8

A. Set up a circuit with a battery, ammeter, and 15 cm length of nichrome wire in series. Connect a voltmeter so that it will measure the voltage across the nichrome wire. Record the current through the resistor and the voltage across it.

B. Connect a second battery in series with the first. Record the current through the resistor and the voltage across it.

C. Connect a third battery in series with the first two and repeat your measurements.

D. What happens to the voltage across the nichrome wire as the number of batteries in the circuit is increased?

   As the voltage across the nichrome wire increases, what happens to the current through the wire?

   How would you compare the results of this experiment with those obtained in Experiment 7.2?

By adding batteries to the circuits in Experiments 7.2 and 7.8, we were able to examine the effect that changing the voltage across a network has on the current through that network. In the following experiments, we change the resistance of an element in a circuit and examine the effect this has on the voltages across and the currents through all the elements in a circuit.

Experiment 7.9

Obtain a 30 cm length of nichrome wire and set up the circuit shown.

A. Measure the voltage across *AB, BC, CD, DE, BD*, and *AE* when there is a 10 cm length of nichrome wire in the circuit. What relationships can you find among these voltages?

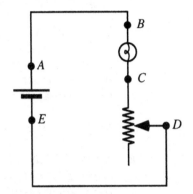

B. With the voltmeter connected across the battery (at points *A* and *E*), vary the current in the circuit by decreasing the length of nichrome wire in the circuit from 30 cm to 0 cm in steps of 10 cm. Record the voltage at each step in a table like the one below.

How does the voltage across the battery vary with increasing current?

What do you observe about the voltage across the battery as the resistance in the circuit is changed?

Length of nichrome wire	Voltage across battery	Voltage across bulb	Voltage across wire
30 cm			
20 cm			
10 cm			
0 cm			

C. With the voltmeter connected across the bulb, vary the current through the bulb by the method of part B. Record your measurements for the voltage across the bulb in your table.

How does the current through the bulb vary with increasing voltage?

D. Repeat the same experiment as in parts B and C, but this time measure the voltage across the nichrome wire. How does the voltage across the nichrome wire vary with increasing resistance in the nichrome wire?

What happens to the voltage across the bulb as the resistance in the rest of the circuit increases?

How does the voltage across the bulb compare with the voltage across the nichrome wire as more resistance is added to the circuit?

How would you summarize the relationships among the voltages in the circuit?

✔ Check your results with a staff member.

McDermott & P.E.G., U.Wash./*Physics by Inquiry*

Experiment 7.10

Obtain a 15 cm piece of nichrome wire, and a
30 cm piece that can serve as a variable resistor.

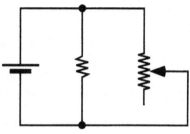

A. Set up the circuit shown at right. Adjust the
variable resistor so there is 10 cm of
nichrome wire in that branch. Predict how
the voltage across the battery will compare
to the voltage across each piece of nichrome wire. Explain your reasoning.

Make the measurements and compare with your predictions.

B. Move the lead of the variable resistor so there is a 30 cm piece of nichrome
wire in the circuit. Predict how the voltage across the battery will compare to
the voltage across each piece of nichrome wire in the circuit. Explain.

Make the measurements and compare with your predictions.

How do the voltages across the two nichrome wires in the circuit
compare to each other and to the voltage across the battery?

Does changing the resistance of a branch connected directly across
a battery significantly affect the voltage across that branch?

Does changing the resistance of a branch connected directly across
a battery significantly affect the voltage across another branch?

Is the voltage across the battery significantly affected by changing
the resistance of any branch connected directly across it?

Batteries in parallel

Thus far, the multiple-battery circuits we have analyzed have only had batteries connected

in series. In the following experiment, we investigate another arrangement of batteries.

Experiment 7.11

A. Set up a circuit consisting of one battery and one bulb. Measure the voltage
across the bulb.

B. Add a second identical battery as shown.

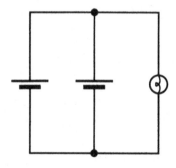

Batteries with their positive terminals connected
together and their negative terminals connected
together are said to be connected in *parallel*.

Observe any changes in brightness in the bulb.
Measure the voltage across the bulb.

McDermott & P.E.G., U.Wash./*Physics by Inquiry*

C. Add a third battery in parallel with the first two batteries. Again, observe any changes in brightness in the bulb. Measure the voltage across the bulb.

What can you conclude about how the current through the bulb changes when additional batteries are added in parallel?

What can you conclude about how the current through each battery changes when an additional battery is added in parallel?

What can you conclude about how the voltage across the bulb changes when additional batteries are added in parallel?

How does the effect of adding batteries in parallel differ from the effect of adding batteries in series?

In Experiments 7.9, 7.10, and 7.11, we observed one of the most important properties of a battery—that it always has approximately the same voltage across its terminals. If a battery is in good condition, the voltage across it does not change appreciably when changes are made in the arrangement, number, and resistance of circuit elements. (Sometimes, when a battery is old, the voltage across the terminals may decrease as the amount of current through the battery becomes large.) For the purposes of the development of our model, we may think of a battery as a constant voltage source.

In Experiments 7.6 and 7.9, we found that the sum of the voltages across all the resistances connected in series with the battery is equal, or almost equal, to the voltage across the battery. As additional resistance is added in series with the original resistance in a circuit, the voltage across the original resistance decreases. However, the sum of the voltages across the old and new parts of the circuit remains equal to the constant, or almost constant, voltage across the battery.

In Experiment 7.10, we found that when resistances are connected in parallel across a battery, the voltage across each branch is equal (or almost equal) to the voltage across the battery. This is true regardless of the size of the resistance in any branch.

Analysis of electric circuits

In the next set of exercises, we examine how the concepts of current and voltage can be

used to predict and explain the behavior of simple circuits. We will find that some

circuits are more easily analyzed in terms of the currents through the circuit elements

while others are more readily understood in terms of the voltages across the elements.

Exercise 7.12

In this exercise we will express mathematically the relationships of current
and voltage in parallel networks.

A. For the parallel network shown, let V_1 be the voltage across
resistor 1, let V_2 be the voltage across resistor 2, and let V_0
be the voltage across the whole network (from A to B).
Write equations to express the relations among these three
voltages.

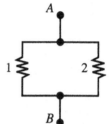

B. For the parallel network shown let i_1 be the current through
resistor 1, let i_2 be the current through resistor 2, and let i_0
be the current through the whole network (entering at A
and leaving at B). Write equations to express the relations
among these three currents.

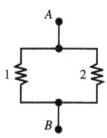

Exercise 7.13

In this exercise we will **express mathematically the relationships of**
currents and voltages in **series networks.**

A. For the series network shown let V_1 be the voltage across resistor 1, let
V_2 be the voltage across resistor 2, and let V_0 be the voltage across the
whole network (from A to B). Write equations to express the relations
among these three voltages.

B. For the series network shown let i_1 be the current through resistor 1, let
i_2 be the current through resistor 2, and let i_0 be the current through the
whole network (entering at A and leaving at B). Write equations to
express the relations among these three currents.

✔ Discuss this exercise and Exercise 7.12 with a staff member.

McDermott & P.E.G., U.Wash./*Physics by Inquiry*

Exercise 7.14

A. Refer to the definitions for series and parallel connections of elements given in Section 4. The definition for a series connection of elements was given in terms of the current passing through the elements. In Exercise 7.13, we found this leads to a simple mathematical relationship between the current through each element connected in series. What is this relationship?

B. In Exercise 7.12, we found there exists a simple mathematical relationship between the voltages across elements connected in parallel. What is this relationship? This relationship is characteristic of parallel branches in circuits. It is often a less cumbersome way of describing parallel connections than that given in Section 4.

As Exercises 7.12 and 7.13 show, the voltages across the elements in series and parallel networks are related to each other differently from the way the currents in these networks are related. We do not speak of voltage, as we do of current, as passing through elements in a circuit. A battery will always have approximately the same voltage across it but does not deliver the same current to all circuits. The following exercises provide practice in thinking about voltage and current in some familiar circuits.

Exercise 7.15

In this exercise, three students give predictions and explanations for the relative brightness of bulbs A, B, and C.

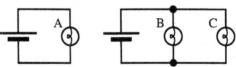

Identify which of the students, if any, are reasoning incorrectly, and determine what is wrong with their reasoning.

Student 1: *"B and C will be dimmer than A. Bulb A gets all of the current from the battery but B and C have to share it."*

Student 2: *"A, B, and C will all be equally bright. They each have the same voltage across them."*

Student 3: *"A, B, and C will all be equally bright. Each has the same resistance, and each is connected directly across the battery, so each bulb has the same amount of current through it. So they are equally bright."*

✔ Check your reasoning with a staff member.

Exercise 7.16

In this exercise, three students give predictions and explanations for the relative brightness of bulbs A, B, and C. Identify which, if any, of the students are reasoning incorrectly, and determine what is wrong with their reasoning.

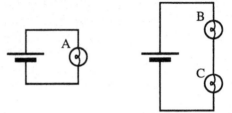

Student 1: *"B and C are equally bright but dimmer than A. B and C have to share the current whereas A gets all of it."*

Student 2: *"B and C are equally bright but dimmer than A. B and C have to share the battery voltage whereas A gets all of it."*

Student 3: *"A is brighter than B and B is brighter than C. B uses up some of the current so less gets through to C. A gets all the current so it is the brightest."*

✔ Check your reasoning with a staff member.

Section 8. Kirchhoff's second rule

In the preceding section we observed that batteries in a complete circuit produce voltages
across the resistances in the circuit. We found that for a two-element series circuit, the
sum of the voltages is equal to the voltage across the battery. For a circuit in which two
elements are connected in parallel directly across the battery, we found that the voltage
across each element is equal to the voltage across the battery. We also noted that we may
regard the battery as a constant voltage source. In this section, we will extend our results
for two-element circuits to more complicated electric circuits.

Experiment 8.1

Set up the circuit shown.

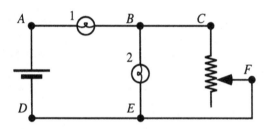

A. Measure the voltage across *AD, AB,
   BE,* and *CF* with lengths of 0, 10,
   20, and 30 cm of nichrome wire in
   the circuit. Record your
   measurements in a table like the
   one below.

Length of nichrome wire	Voltage across battery	Voltage across bulb 1	Voltage across bulb 2	Voltage across wire
0 cm				
10 cm				
20 cm				
30 cm				

B. How does the brightness of bulb 1 vary as the voltage across it
   increases?

C. For each particular length of wire can you find a mathematical
   relationship between the voltages in the circuit?

    McDermott & P.E.G., U.Wash./*Physics by Inquiry*

D. What happens to the resistance of the network of bulb 2 and the wire as the length of nichrome wire in the circuit is increased? What happens to the voltage across this network as its resistance is increased?

E. What happens to the voltage across bulb 1 as the length of nichrome wire in the circuit is increased?

✔ Check your results with a staff member.

We can see from Experiments 7.9 and 8.1 how voltages across series and parallel networks are related. If several elements are connected in series to form a network, the voltage across the network is just the sum of all the voltages across the individual elements. If several elements are connected in parallel to form a network, the voltage across each element and the voltage across the network are the same. These observations lead to a rule for the sum of the voltages across the elements that lie between two nodes in a circuit. This rule is known as *Kirchhoff's second rule:*

The sum of the voltages across all of the elements along any path
is the same for all paths between two nodes.

In the circuit for Experiment 8.1 for example, the voltage across bulb 2 equals the voltage across the nichrome wire. Furthermore, the sum of the voltages across bulb 1 and bulb 2 equals the sum of the voltages across bulb 1 and the nichrome wire. Note that each of these paths can be considered to be in a loop with the battery and that the voltage across each path equals the voltage across the battery. This fact suggests another way of stating Kirchhoff's second rule:

The voltage across the battery in a current loop
is equal to the sum of the voltages across the other elements.

For example, the voltage across the battery in Experiment 8.1 is equal to the sum of the voltages across bulbs 1 and 2 in loop *ABEDA*, and also to the sum of the voltages across bulb 1 and the nichrome wire in loop *ABCFEDA*. The two forms of Kirchhoff's second rule that appear above are equivalent. (Kirchhoff's first rule was discussed in Section 5.)

Exercise 8.2

In this exercise, four students give explanations for the relative brightness of the bulbs in the circuit below.

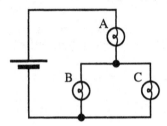

Student 1: "B and C are equally bright but dimmer than A. B and C have to share the current whereas A gets all of it. Therefore A is brighter than B or C."

Student 2: "Bulb A has more resistance than the B-and-C network so bulb A has more voltage across it. Therefore A is brighter than B or C."

Student 3: "Bulb A uses up most of the current so less is left for B and C. A is therefore brighter than B or C."

Student 4: "After bulb A, the voltage divides into two paths with the result that B and C each get less voltage than A. Therefore A is brighter than B or C."

Identify which of the students, if any, are reasoning incorrectly, and determine what is wrong with their reasoning.

Experiment 8.3

In Experiment 3.10, you examined the circuit from Exercise 8.2 and were asked to consider how the bulbs would be affected when bulb C is unscrewed. Recall that it is difficult to answer this question using only the model for *electric current*. Below, we apply the concept of voltage.

A. Consider how the following quantities change when bulb C is unscrewed and explain your reasoning:

- the current through bulb A.

- the voltage across bulb A.

- the voltage across bulb B.

- the current through bulb B.

B. On the basis of your answers above, predict how the brightness of bulb A and bulb B would change when bulb C is unscrewed. Explain.

Set up the circuit and check your predictions.

✔ Check your reasoning with a staff member.

Experiment 8.4

A. Select four identical bulbs by connecting each in series with a battery and an ammeter.

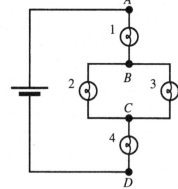

B. Connect the four identical bulbs as shown at right. Measure the voltage V_{bat} across the battery, and the voltage V_1 across bulb 1.

Predict (do not measure) the voltage across bulb 4.

C. Now, measure the voltage V_4. How does your measurement compare with your prediction? Would your answer change if bulbs 1 and 4 were not identical? Explain.

D. Use Kirchhoff's second rule to predict the voltage across terminals B and C. Check your prediction.

Experiment 8.5

The circuit below consists of five identical bulbs and a 1.5 V battery.

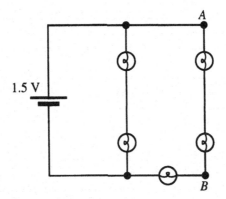

A. Predict the voltage across terminals A and B.

B. Set up this circuit and check your prediction.

Exercise 8.6

In the circuit shown at right, the voltage across resistor 1 is x times as large as the voltage across resistor 2. Write expressions for the voltage across each resistor in terms of x and the battery voltage.

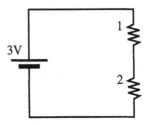

Multiple batteries in a single current loop

So far we have only considered series connections of batteries in which the positive terminal of one battery is connected to the negative terminal of the other. In this case, we found that the total voltage across the battery combination was equal to the sum of the two battery voltages. There are other ways to connect batteries in series. The positive terminals of the two batteries can be connected to each other or the two negative terminals can be connected to each other. In addition, the batteries in a current loop need not directly follow one another but can have another element between them in the circuit. In the following experiments, we examine these more general multiple battery circuits.

Experiment 8.7

Obtain a 15 cm length of nichrome wire, a bulb, and two identical batteries.

A. Set up circuit 1 as shown at right. Predict how the voltage across the battery combination will compare to the total voltage across the circuit elements.

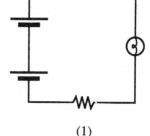

B. Measure the voltage across each battery (V_{bat1}, V_{bat2}), the battery combination (V_{bat}), the bulb (V_{bulb}), and the nichrome wire (V_{wire}).
Check your predictions.

(1)

How does the voltage across the battery combination compare to the voltage across the individual batteries? Write an algebraic equation relating these quantities.

How does the voltage across the battery combination compare to the voltage across the elements in the circuit? Use Kirchhoff's second rule to write an algebraic equation relating these quantities.

Write an algebraic equation relating the voltages V_{bat1}, V_{bat2}, V_{bulb}, and V_{wire}.

C. Set up circuit 2 as shown at right. Measure the voltage across each battery, the bulb, and the nichrome wire.

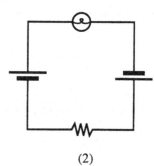

How does the voltage across each element in circuit 2 compare to the voltage across the same element in circuit 1?

Write an algebraic equation relating the voltages V_{bat1}, V_{bat2}, V_{bulb}, and V_{wire}.

(2)

Do the locations of the batteries in a single loop circuit affect the voltage across any of the elements?

Use the results of this experiment to generalize Kirchhoff's second rule so that it applies to circuits with one or more batteries not necessarily connected one right after the other.

✔ Check your results with a staff member.

Experiment 8.8

A. Set up a circuit as shown at right, so that the batteries are connected at their negative terminals. Predict how the voltage across the battery combination compares to the voltage across the bulb.

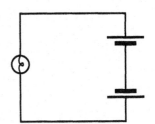

Measure the voltage across the bulb (V_{bulb}), each battery (V_{bat1}, V_{bat2}), and across the battery combination (V_{bat}).

How is the voltage across the battery combination related to the voltage across the bulb? Write an equation expressing this relationship.

How is the voltage across the battery combination related to the voltage across the individual batteries? Write an equation expressing this relationship. Does one battery tend to reinforce or cancel the effect of the other?

Write an equation that relates the voltages V_{bulb}, V_{bat1}, and V_{bat2}.

Do you think it would make a difference if there were a resistance connected between the batteries?

B. Repeat this experiment for the case in which the positive terminals of the batteries are connected together.

C. Use the results of this experiment to write Kirchhoff's second rule in a more general form to include all arrangements of batteries in a single loop circuit.

✔ Check your results with a staff member.

McDermott & P.E.G., U.Wash./*Physics by Inquiry*

Exercise 8.9

Consider the following circuit.

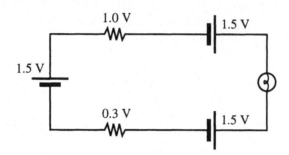

Use your modified version of Kirchhoff's second rule to find the voltage across the bulb.

Experiment 8.10

A. Five different circuits can be formed with two batteries and one bulb. Since one of these circuits wears out batteries very quickly, do not set up the circuits yet. Instead, draw a diagram for each of the five circuits. Use your model for electric current and Kirchhoff's second rule to predict what will happen in each circuit. Check your predictions with a staff member.

B. Set up the circuits that will not damage the batteries and test your predictions for these circuits.

C. Measure the voltage across the bulb, each battery, and the battery combinations in each of the safe circuits. Compare the voltage across the bulb in these circuits. Also, for each circuit, compare the voltage across the bulb with the voltage across the battery combination.

D. Compare the brightness of the bulb in each circuit.

The voltmeter as a circuit tester

When a circuit does not behave as intended, the problem is often an open circuit or a short circuit. A voltmeter is an excellent instrument for detecting either of these defects because it can be used to test a circuit without disconnecting any of the elements.

The following two experiments show the patterns of voltages that are associated with these two common types of electrical malfunctions.

Experiment 8.11

Set up the circuit shown below and measure the voltage across the battery, each bulb, and the switch. Measure the voltages both when the switch is open and when the switch is closed.

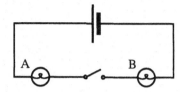

Experiment 8.12

Set up the circuit shown below and measure the voltage across the battery, each bulb, and the switch. Measure the voltages both when the switch is open and when the switch is closed.

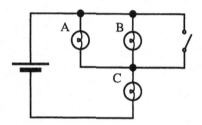

As Experiment 8.11 indicates, an open circuit is characterized by zero voltage across the good elements in series with the break. All of the voltage across the battery is indicated on the voltmeter when it is connected across the break in the circuit. As Experiment 8.12 indicates, a short circuit is characterized by zero voltage across the shorted elements. Other elements still show a non-zero voltage across them.

If a circuit has several defects, many tests may be required to find them all. The tests for open and short circuits described above suffice only when a single defect is present.

Experiment 8.13

Ask a staff member to give you the circuit for this experiment. When the switch is closed, all of the bulbs should light, but they do not. Find what is wrong by using only a voltmeter and without removing or replacing any of the circuit elements. Do not disconnect any wires or form any new circuits. You may remove bulbs from their sockets, but you may not use equipment other than that which is provided.

Experiment 8.14

Ask a staff member to give you the circuit for this experiment. Find out what is wrong with it following the same rules as in Experiment 8.13.

Experiment 8.15

Ask a staff member to give you the circuit for this experiment. Find out what is wrong with it following the same rules as in Experiment 8.13.

Experiment 8.16

Ask a staff member to give you the circuit for this experiment. Find out what is wrong with it following the same rules as in Experiment 8.13.

Exercise 8.17

Ask a staff member to show you the circuit for this exercise. Some of the bulbs in this circuit are not lit.

How can you determine whether there is a defect in this circuit, or whether the bulbs are lit too dimly to see? If there is a defect, how can you locate it?

Use the method you have described to locate any defects that may be present in this circuit.

Draw a standard circuit diagram and decompose the circuit into its series and parallel networks. Can you identify the series and parallel networks in the actual circuit?

Exercise 8.18

Describe a way to use a bulb, instead of a voltmeter, to test for defects in circuits.

Section 9. Series and parallel decomposition

We have developed procedures that enable us to make predictions about the behavior of
simple circuits consisting of either series or parallel combinations of resistances. Many
circuits that appear at first to be too complicated for this type of analysis can be
decomposed into series and parallel combinations of elements and networks.

Sample problem

Analyze by series and parallel
decomposition the circuit at right
consisting of one battery and seven
resistors.

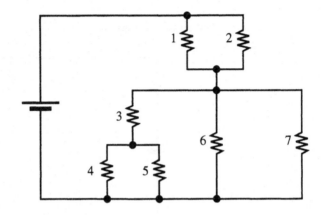

Sample solution

Step 1: Identify two or more single elements that are connected in series or in parallel.

No two of the resistors are in series with each other. Resistors 1 and 2 are
in parallel, as are resistors 4 and 5. Resistors 6 and 7 are also in parallel.

It is usually best to start with the elements deepest
within the circuit and work outward. In this case,
we will start with resistors 4 and 5. The network
consisting of these two elements connected in parallel
will now be treated as a single entity called network A.

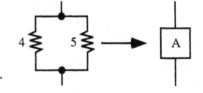

The circuit diagram now becomes:

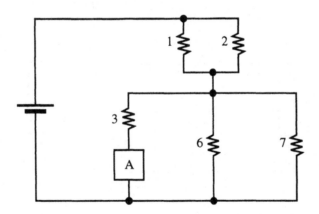

Step 2: We can now see that network A and resistor 3 are connected in series. The network consisting of these two elements in series can then be treated as a single entity called network B.

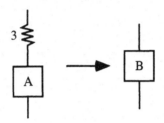

The circuit diagram now becomes:

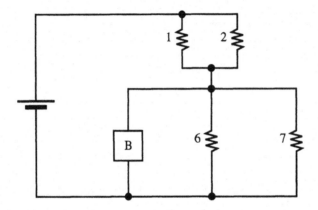

Step 3: Since resistors 1 and 2 are connected in parallel, we will treat them as a single entity named network C. In addition, network B and resistors 6 and 7 are all connected in parallel. The network consisting of these three elements will be called network D.

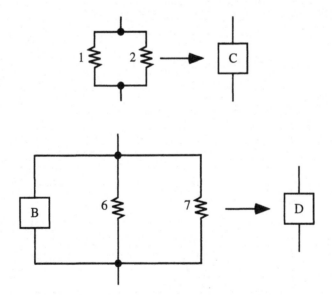

The circuit diagram now becomes:

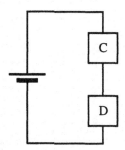

Networks C and D are connected in series.

Sample problem

To show how this kind of decomposition can be helpful in analyzing a circuit, we will pose two sample questions:

(1) If the resistance of resistor 1 is increased, what should be done to resistor 2 to keep the currents in the rest of the circuit the same?

(2) What happens to the current in resistor 1 if resistor 4 is removed from the circuit by disconnecting it?

Sample solution

To answer question 1, we look at the following representation of the circuit:

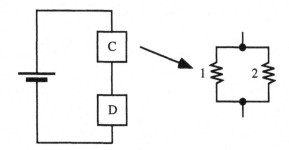

Since networks C and D are in series, we can keep all of the currents in network D the same if we just make sure that the current through network C does not change. Since we have increased the resistance of resistor 1, we must decrease the resistance of resistor 2 in order to let through the extra current that no longer passes through resistor 1.

McDermott & P.E.G., U.Wash./*Physics by Inquiry*

Question 2 can be answered by the following chain of reasoning. If resistor 4 is disconnected, the resistance of network A will be larger, and therefore the resistance of network B will be larger.

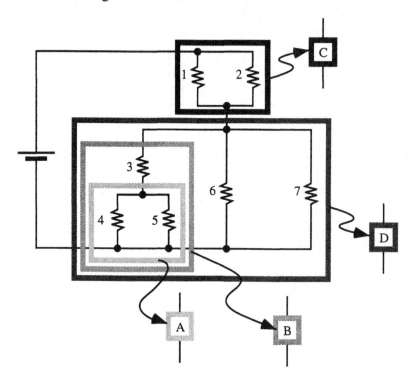

The resistance of network D will therefore be larger, and there will be less current through the series circuit of networks C and D. Since resistor 1 is a part of network C, it will have less current through it as well.

Exercise 9.1

Decompose the following circuit into its series and parallel networks. Is it basically a series circuit or a parallel circuit?

Suppose the resistance of resistor 5 is increased.

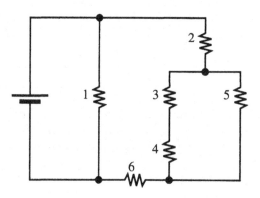

(1) What happens to the current through resistor 1?

(2) What happens to the current through the battery?

✔ Check your results with a staff member.

Exercise 9.2

Decompose the following circuit into its series and parallel networks. Is it basically a series circuit or a parallel circuit?

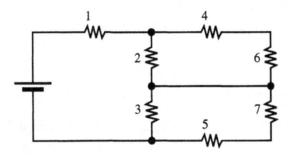

What happens to the current through resistor 3 if resistor 4 is replaced by a connecting wire?

Exercise 9.3

Ask a staff member to show you the circuit for this exercise. Draw the diagram and decompose the circuit into its series and parallel networks. Can you identify the series and parallel networks on the actual circuit?

Exercise 9.4

Consider the following discussion about the circuit shown below.

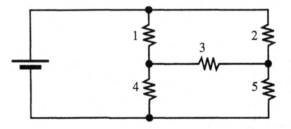

Student 1: *"I can't decompose this circuit into series and parallel parts."*

Student 2: *"You have to take it step by step. 1 and 2 are in parallel. Let's call that A. 4 and 5 are in parallel. Let's call that B. Then A and 3 and B are one after another in series."*

Student 2 is not correct. Describe the errors that student 2 has made.

McDermott & P.E.G., U.Wash./*Physics by Inquiry*

In this section, we have developed a method for decomposing complicated circuits into series and parallel combinations of elements and networks. The method involves identifying a series or parallel network of elements within the circuit and treating that network as a single entity having two terminals (two connections to the rest of the circuit).

The circuit in the preceding exercise cannot be classified as either series or parallel. Although at first glance the resistors in the network may appear to be connected in parallel, more careful inspection reveals they are not. No two elements in that circuit may be treated as a single entity with only two terminals. For example, an entity representing resistors 1 and 4 would have three terminals (three connections to the rest of the circuit). We therefore cannot classify that connection of elements as either series or parallel.

It is not only the connections within a network that are important in deciding whether the network is series, parallel, or neither. The connections to the rest of the circuit are also important.

Exercise 9.5

Consider the following circuits. Note that the network consisting of five resistors is identical in each circuit.

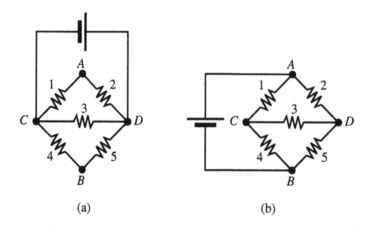

(a)   (b)

For each circuit, describe the series and parallel connections of resistors. If the resistors are not connected in series or in parallel, state that explicitly. Explain your reasoning.

Circuits not reducible to series and parallel combinations

Some circuits cannot be entirely decomposed into series and parallel networks. The circuit in Exercise 9.4 is an example. None of the resistors is connected in series or in parallel with any other. This particular circuit is called a *bridge circuit.*

A key step in analyzing voltages in bridge circuits and others that cannot be reduced to simple series or parallel combinations is to measure the voltages across successive elements along a path from one node to another. This path is not necessarily the only path followed by the current in the circuit but is merely a continuous part of the circuit between the nodes.

Examine the diamond-shaped circuit below *(ACDB).* An electrical element (resistor 2) bridges two branches of the circuit so that it cannot be reduced to series and parallel combinations.

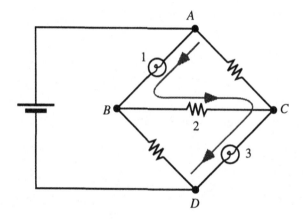

Consider the particular path from *A* to *D* that passes through bulb 1, resistor 2, and bulb 3. If we were to trace this path measuring voltages, we would first measure the voltage across *AB,* then *BC,* then *CD.* The sum of these three measurements would be called the sum of the voltages along path *ABCD.*

When summing the voltages along a path, we must be sure always to place the positive terminal of the voltmeter closer to the positive terminal of the battery and the negative terminal of the voltmeter farther along the path. We will clearly obtain positive measurements for bulbs 1 and 3. The voltmeter may, however, have a negative deflection for an element like resistor 2 whose "positive" side is not obvious. If a negative

deflection of the meter is obtained, the leads must be reversed to obtain a measurement.

The voltage across the resistor must be considered negative in this case and should

appear in the sum of the voltages as a negative number.

Experiment 9.6

Set up the bridge circuit shown with two bulbs and three 10 cm pieces of nichrome wire as shown.

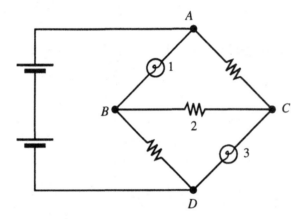

A. Find the four paths from node *A* to node *D*.

B. Determine the sum of the voltages across all of the circuit elements along each path.

Compare the sums of the voltages along the different paths.

C. Identify, if possible, the direction of the current in each of the circuit elements.

D. The voltage across one of the circuit elements should be negative. What does the negative sign tell us?

✔ Check your results with a staff member.

Exercise 9.7

In the circuit below, all but two of the voltages have been measured and appear in the diagram. (The + and − signs indicate to which side of each element the positive and negative terminals of the voltmeter were connected to give the labeled voltage across that element.)

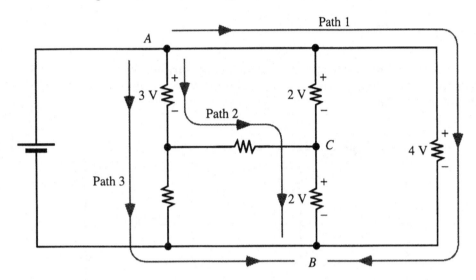

Three paths from node A to node B are shown. Determine the sum of the voltages across all the circuits elements along each of the three paths from node A to node B:

- Path 1: Write the successive terms in the sum of voltages across all the circuit elements starting at node A. What is the sum of the voltages along this path?

- Path 2: Repeat the procedure above for path 2 starting at node A. Include the unknown voltage in your sum and determine its value. Explain your reasoning.

- Path 3: Repeat the procedure above for path 3 starting at node A.

In Experiment 9.6 and Exercise 9.7, we have illustrated how Kirchhoff's second rule can be applied to circuits that cannot be decomposed into series and parallel networks. In such circuits, it is often difficult to predict the direction of the current in all the elements. In solving problems using Kirchhoff's second rule, the directions of the paths from one node to another are arbitrary choices that we make and may not correspond to the actual direction of the current. We obtain the correct answer as long as we are consistent in assigning signs and in interpreting our results. If the answer we obtain for the voltage across a resistor is positive, the correct choice of direction has been made. A negative

voltage across a resistor means that the voltages have been summed along a path through

the resistor that is in the opposite direction to the current.

Exercise 9.8

The circuit below consists of nine resistors of various sizes and one 12 V battery. The voltages across some of the resistors are given. Find the remaining voltages.

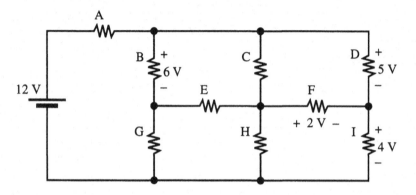

Exercise 9.9

In the bridge circuit at right, the voltages V_1, V_2, and V_3 are to be determined. It is known that V_3 is equal to the sum of voltages V_1 and V_2. Use this information along with Kirchhoff's second rule to find the three voltages.

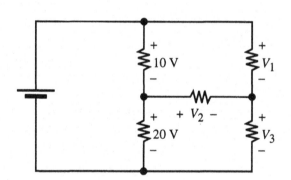

Exercise 9.10

The circuit at right consists of six identical bulbs and one 14 V battery. The voltages across two of the bulbs are given.

Find the remaining voltages.

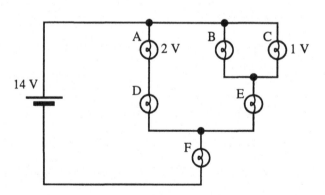

Section 10. Ohm's law

In this section, we examine quantitatively how current, voltage, and resistance are related to one another. We extend our model to include a rule for finding numerical values for these variables in circuits in which the resistance is due only to linear resistors.

In Experiment 3.1, we found that adding bulbs in series with an indicator bulb results in a dimming of the indicator bulb. We assumed that the dimming of the bulb is the result of a decrease in current through the bulb. By thinking of the bulbs as obstacles to the current, we were then able to account for the dimming (i.e., adding bulbs in series increases the obstacle to the current, thus decreasing the flow).

In Experiment 7.1, we increased the obstacle to the flow in a circuit by adding a bulb in series with an indicator bulb. We then found that adding a battery in series restored the indicator bulb to its previous brightness. These observations are consistent with the assumption that the battery is the agent that "pushes" current through a circuit. The greater the obstacle presented by an element, the less current "pushed" through that element by the battery.

We can determine the size of the obstacle presented by an element by finding the amount of "push" required to cause a unit of current to flow through that element. In other words, we want to find the number of volts required for an ampere to flow through the element.

Experiment 10.1

A. Set up a circuit with a battery and a 40 cm piece of nichrome wire in series with an open switch. Draw a circuit diagram that shows how you would connect an ammeter to measure the current, i, through the resistor and a voltmeter to measure the voltage, V, across the resistor. Check your diagram with a staff member.

Measure the current and voltage. Be sure to take the necessary precautions to protect the meters. (See Sections 5 and 7.)

B. Make a table for recording measurements for the current through the resistor as the voltage across the resistor is varied (by adding batteries in series). Record your measurements from part A in the table.

Add a second battery in series with the first. Record the values for the voltage and current.

Add a third battery in series with the first two. Record your measurements.

C. Plot the data from your table on a graph of voltage versus current. Label the vertical axis voltage, V, and the horizontal axis current, i.

Should your graph go through the origin? Explain.

Is the graph straight or curved? How would you characterize the relationship between the voltage and current for this resistor? Are the voltage and current directly proportional, inversely proportional, or neither?

Interpret the slope of your graph at a particular point. Is the slope the same or different for different points?

D. Predict how the graph in part C would look if you had used a 25 cm piece of nichrome wire instead of a 40 cm piece of nichrome wire.

How would the graph look if you had used a 55 cm piece of nichrome wire?

Check your predictions.

Experiment 10.2

A. Repeat Experiment 10.1 using a bulb instead of nichrome wire.

B. Contrast the behavior of the nichrome wire and the bulb as the current in the circuit varies.

In Experiment 6.5 you examined how adding bulbs in series affected the equivalent resistance of a network. Review your conclusions from part C of that experiment. Are your observations from that experiment consistent with your results from part A above? Explain.

C. We have followed the common practice of using the term "resistor" instead of "linear resistor" for nichrome wire. Explain why nichrome wire is more properly called a linear resistor.

✔ Discuss your results with a staff member.

In Experiment 10.1, we found that for nichrome wire, the ratio *V/i* is constant. This ratio is approximately constant for any linear resistor over a wide range of currents and voltages. In Experiment 10.2, we found that for bulbs this ratio is not constant but increases as the voltage across the element is increased.

We have thought of bulbs and resistors as presenting an obstacle, or resistance, to current. We can now make this idea more precise. For nichrome wire and other linear resistors, we can define the *resistance* as the constant ratio of the voltage across the resistor to the current through the resistor *(V/i)*. We interpret the result of the division as the number of volts required to drive one ampere of current through the resistor. The unit of resistance (volt per ampere) is called the *ohm,* which is abbreviated with a capital omega, Ω. For example, a resistance of one ohm means that for one ampere of current to flow through that resistance, one volt must be applied across the terminals of the resistor by the battery.

Exercise 10.3

When a certain linear resistor has 90 V across it, there are 15 A through it.

A. How many volts are required to drive 10 A through the same resistor?

B. What is the current through the resistor if the voltage across it is changed to 33 V?

C. If the voltage across the resistor is doubled, what happens to the current through it?

Explain your reasoning.

Exercise 10.4

A. Two different linear resistors, A and B, have the same current through them. If the voltages across them are 5 V and 4 V, respectively, what is the ratio of the resistances?

B. Two different linear resistors, A and B, have the same voltage across them. If the currents through the resistors are 5 A and 4 A, respectively, what is the ratio of the resistances?

Explain your reasoning.

The relationship you observed in Experiment 10.1 between the current and voltage for the nichrome wire is referred to as *Ohm's law*. Ohm's law can be expressed in the form of an equation as follows:

$$R = \frac{V}{i} = \text{constant (for linear resistors only)}$$

Ohm's law is frequently expressed in the following equivalent forms:

$$i = \frac{V}{R}$$

$$V = iR$$

Ohm's law is qualitatively different from Kirchhoff's rules. Kirchhoff's rules are fundamental statements about the nature of current and voltage. These rules are believed to be exactly true. Ohm's law, however, is never exactly true and is not viewed as a fundamental relation. It is almost true for nichrome wire, copper wire, and commercially produced resistors, but is far from true for bulbs and many other kinds of circuit elements.

We will now analyze some circuits with linear resistors by using Ohm's law and Kirchhoff's two rules. The first two examples, when applied to networks as well as single resistors, illustrate basic techniques that can be used to solve many circuit problems.

Sample problem: the current divider

A total current i_0 flows into a parallel network of resistors R_1 and R_2. What is the current, i_1, through the first resistor and the current, i_2, through the second resistor?

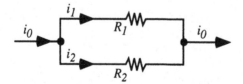

Sample solution

Step 1: To solve this problem we need both of Kirchhoff's rules and Ohm's law. Kirchhoff's first rule gives:

$$i_0 = i_1 + i_2$$

Kirchhoff's second rule requires that the voltages, V_1 and V_2, across the two resistors be equal. We will call them simply V.

$$V = V_1 = V_2$$

Ohm's law applied separately to each resistor gives two equations:

$$V_1 = i_1 R_1 \quad \text{and} \quad V_2 = i_2 R_2$$

Step 2: Eliminating V from these last two equations we get:

$$i_1 R_1 = i_2 R_2$$

or

$$\frac{i_2}{i_1} = \frac{R_1}{R_2}$$

When two resistors are connected in parallel,
the ratio of the currents is the inverse of the ratio of the resistances.

For example, if a 10-ohm resistor and a 20-ohm resistor are connected in parallel, the 10-ohm resistor will have twice as much current through it as the 20-ohm resistor.

Note: It is *not* true that "current takes the easiest path" exclusively. There is always some current through each branch in inverse proportion to the resistance of that branch.

Step 3: To complete the problem, we solve for each current in terms of the total current. From equations in steps 1 and 2, we get:

$$i_1 = \frac{R_2}{R_1 + R_2} i_0$$

$$i_2 = \frac{R_1}{R_1 + R_2} i_0$$

Exercise 10.5

Three resistors with resistance 4 ohms, 2 ohms, and R are connected in parallel. There is a total current i through the network. Find the current through each resistor in terms of R and i. Do *not* try to use the results from the sample problem; start from Kirchhoff's two rules and Ohm's law.

Sample problem: the voltage divider

Find the voltage V_1 across the first resistor and the voltage V_2 across the second resistor in the circuit below.

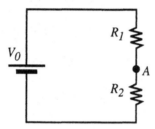

Sample solution

Current conservation at node A requires the currents through the two resistors to be the same. The equality of the currents is the key to the solution of this problem.

Step 1: We will now write several equations:

$$V_0 = V_1 + V_2$$ The sum of the voltages across the resistors must equal the voltage across the battery (Kirchhoff's second rule).

$$V_1 = iR_1$$ Ohm's law for the first resistor.

$$V_2 = iR_2$$ Ohm's law for the second resistor.

Step 2: Dividing the two Ohm's law equations, we get:

$$\frac{V_2}{V_1} = \frac{iR_2}{iR_1} = \frac{R_2}{R_1}$$

When two resistors are connected in series, the voltages across the resistors are in the same ratio as the ratio of the resistances. For example, if a 10-ohm resistor and a 20 ohm resistor are connected in series, the voltage across the 10-ohm resistor will be half the voltage across the 20-ohm resistor. The total voltage across both resistors is divided among the resistors in proportion to their resistance.

We can now solve for the voltages across resistor 1 and resistor 2. From the equations in steps 1 and 2:

$$V_1 = \frac{R_1}{R_1 + R_2} \; V_0$$

$$V_2 = \frac{R_2}{R_1 + R_2} \; V_0$$

Exercise 10.6

Find the voltage across each resistor in the circuit below in terms of R. Do *not* try to use the results from the sample problem; start from Kirchhoff's two rules and Ohm's law.

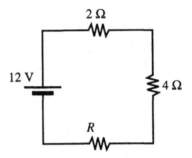

Series and parallel reduction of networks of resistors

Much of the algebra involved in solving circuit problems can be eliminated by

decomposing a circuit into its series and parallel parts and replacing each part with an

"equivalent resistance." In Section 6, you found the equivalent resistance of series and

parallel networks by measuring an equivalent length of nichrome wire. That is, you

found the length of nichrome wire that represents the same obstacle as the network.

Below, we develop a mathematical derivation of the rules for calculating the equivalent

resistance.

Series networks

In a series network, the current is the same in all elements, and the voltages across the elements add to give the total voltage across the network.

The voltage across the entire network is the sum of the individual voltages across the elements:

$$V_t = V_1 + V_2 + V_3 + ...$$

There is the same current through all the resistors. Therefore, all of the individual currents can be represented by the same variable, i, which will also be the total current through the network.

$$i = i_1 = i_2 = i_3 = ...$$

If we use Ohm's law to express each of the voltages above in terms of the common current, we get:

$$iR_t = iR_1 + iR_2 + iR_3 + ...$$

where R_t is the total resistance of the entire network, sometimes called "the equivalent resistance of the network."

Dividing by i, we get a rule for combining resistances in series:

$$\text{SERIES NETWORKS:} \quad R_t = R_1 + R_2 + R_3 + ...$$

Parallel networks

In a parallel network, the voltage is the same across all the elements, and the currents add to give the total current through the network.

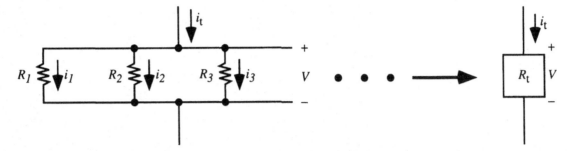

The current, i_t, through the network as a whole is the sum of the individual currents through the individual elements:

$$i_t = i_1 + i_2 + i_3 + \ldots$$

All of the resistors have the same voltage across them. Therefore, all of the individual voltages can be represented by the same variable, V, which can also represent the voltage across the network as a whole.

$$V = V_1 = V_2 = V_3 = \ldots$$

If we use Ohm's law to express each of the currents above in terms of the common voltage across the parallel elements, we get:

$$\frac{V}{R_t} = \frac{V}{R_1} + \frac{V}{R_2} + \frac{V}{R_3} \ldots$$

Dividing by V, we get a rule for combining resistances in parallel:

$$\text{PARALLEL NETWORKS}: \quad \frac{1}{R_t} = \frac{1}{R_1} + \frac{1}{R_2} + \frac{1}{R_3} \ldots$$

When used in combination with the technique of decomposing a circuit into series and parallel networks, the formulas for equivalent resistance of series and parallel networks can simplify the analysis of many circuits.

Exercise 10.7

Are the rules developed in this section for calculating the equivalent resistance consistent with the rules you found by experiment in Section 6? Explain.

Sample problem

Find all of the currents through and voltages across the resistors in the following circuit:

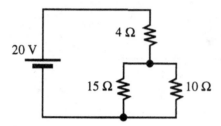

Sample solution

First, replace the two resistors in parallel with an equivalent resistor. The equivalent resistance is found from:

$$\frac{1}{R} = \frac{1}{10\,\Omega} + \frac{1}{15\,\Omega}$$

This gives $R = 6$ ohms.

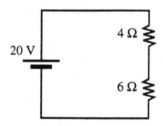

We now have resistances of 4 ohms and 6 ohms in series.

As we found for the voltage divider in the sample problem earlier in this section, the voltages across these resistors will be in the same ratio as the resistances, 4/6. Thus the voltage across the 4-ohm resistor will be 4/10 of the battery voltage, 20 V × 4/10 = 8 V. The voltage across the 6-ohm network will be the remaining 20 V − 8 V = 12 V.

Ohm's law now gives the currents through each of the resistors:

$$\frac{12\,V}{10\,\Omega} \quad = \quad 1.2\,A \text{ through the 10 ohm resistor}$$

$$\frac{12\,V}{15\,\Omega} \quad = \quad 0.8\,A \text{ through the 15 ohm resistor}$$

$$\frac{8\,V}{4\,\Omega} \quad = \quad 2.0\,A \text{ through the 4 ohm resistor}$$

Exercise 10.8

Find the current through the battery in the following circuit of nine 5-ohm resistors and one 20 V battery.

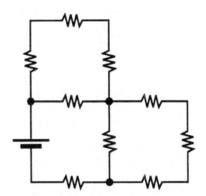

Exercise 10.9

Find the voltage between nodes A and B in the circuit below.

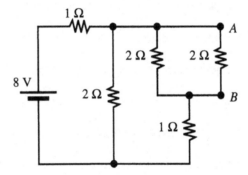

Exercise 10.10

Find the current through the 300-ohm resistor between nodes A and B in the circuit below.

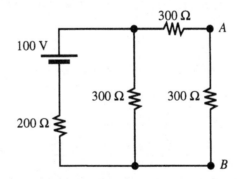

Exercise 10.11

Find the voltage across the 40-ohm resistor and the current through the 10-ohm resistor in the circuit below.

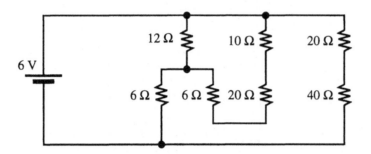

The decomposition of a circuit into its series and parallel connections is a labor-saving convenience, not a necessity. It is often helpful to analyze parts of a circuit in this way, but it is rarely necessary to reduce the entire circuit to a single equivalent resistance. The preceding exercise is a case in point. This problem is solved most easily by using the independence of parallel branches connected directly across the battery.

In many circuits, it is not necessary to solve for any of the currents if the quantity desired is a voltage across a particular resistor. Two rules help us work with voltages alone. The first is that the voltages across several resistors or networks connected in parallel are the same. The second is that the voltages across resistors in series are in the same ratio as the resistances.

Exercise 10.12

Find all of the voltages across the resistors below without calculating any currents.

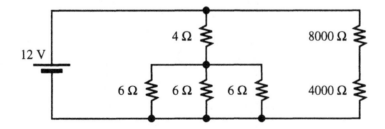

Exercise 10.13

Find all of the voltages across the resistors below without calculating any currents.

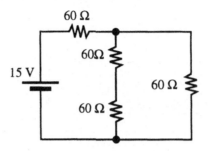

Exercise 10.14

Find all of the voltages across the identical resistors below without calculating any currents.

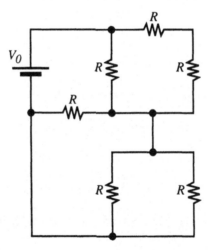

Part D: Batteries and bulbs in everyday life

Thus far in *Electric Circuits*, we have assumed that the batteries we have used are *ideal,* and that the bulbs we have used are identical. In part D, we extend our model to account for the behavior of real batteries and non-identical bulbs.

Section 11. Real batteries

Experiment 11.1

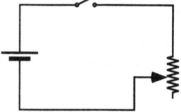

Set up the circuit shown at right with a variable resistor consisting of about 30 cm of nichrome wire.

Record the voltage across the battery with lengths of 30 cm, 20 cm, and 10 cm in the circuit. Also record the voltage across the battery with the switch open.

One of the assumptions we have been making in our analysis of circuits is that the voltage is always the same across a given battery. However, this is only approximately true. As the previous experiment illustrates, the voltage measured across a battery may change when there is a large current through the battery.

When there is no current through a battery, the voltmeter reading is called the *open-circuit voltage.* Usually the difference between the actual voltage (the voltage you would measure) and the open-circuit voltage is small. However, as the current through the battery increases, this difference may become large. This effect becomes more pronounced as a battery ages.

Internal resistance

The change in voltage across a battery when the current through it is changed may be thought of as being a result of resistance within the battery, called *internal resistance.* In a real battery, the internal resistance varies considerably during its lifetime. For a new

 McDermott & P.E.G., U.Wash./*Physics by Inquiry*

battery, the internal resistance may be very low. As a battery is used, however, or even if it stays on the shelf, its internal resistance gradually increases.

A real battery can be thought of as consisting of a constant resistor in series with an ideal battery. Whenever internal resistance is not negligible, we will represent a battery on a circuit diagram by two elements: an ideal battery and an internal resistance r. The following drawing illustrates this representation.

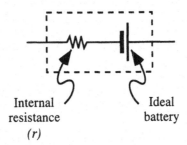

Internal
resistance
(r)

Ideal
battery

The dashed line, which symbolizes the entire (real) battery, is not always drawn on a circuit diagram. Note that we are now refining our definition of the symbol

to represent only the ideal part of the battery. The voltage across the ideal part of the battery is always equal to the open-circuit voltage of the real battery. However, the reading on a voltmeter placed across the terminals of the battery would include both the voltage across the ideal battery and the voltage across the imaginary resistor.

Exercise 11.2

In this exercise we will use the data from Experiment 11.1 to determine the internal resistance of the battery.

A. For each closed circuit in Experiment 11.1, calculate the "missing voltage," where the missing voltage is the open-circuit voltage minus the actual measured voltage.

B. The resistance of 10 cm of 28-gauge nichrome wire is about 1.4 ohms. Calculate the current through the battery for each closed circuit.

C. Plot a graph of the missing voltage versus the current through the battery for each circuit. Include the data from the open circuit on your graph.

      The graph you obtain should be approximately straight. What does this imply about the internal resistance of the battery?

      Find the value for the internal resistance of the battery.

✔ Check your reasoning with a staff member.

The following set of terms is used when the internal resistance is significant:

Open-circuit voltage: the voltage across a battery that has no current through it.

Terminal voltage: the voltage across a battery in an operating (closed) circuit.

Load or *load resistance:*   the part of the circuit that does not include the battery or its internal resistance.

Exercise 11.3

What will be the current through the load resistor in the circuit at the right? The open-circuit voltage of the battery is 8 V.

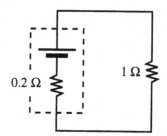

Exercise 11.4

A certain battery has an open-circuit voltage of 6.00 volts. When connected to the load shown in the circuit below, the battery produces a current of 1.00 A in each of the 4-ohm resistors. Find the load resistance and the internal resistance of the battery.

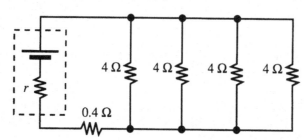

Exercise 11.5

Suppose we have a battery with open-circuit voltage V_0 and internal resistance r. What load resistance will reduce the terminal voltage to exactly half of the open-circuit voltage?

Short circuits

Thus far, we have treated a shorting wire as having little or no resistance. However, in actual practice, a connecting wire may have some resistance. When the resistance of the connecting wire is significant with respect to the resistance of the other elements in the circuit, it is customary to show the connecting wire as a resistor on a circuit diagram. Under these circumstances, the internal resistance of the battery plays a role in determining the current and voltage in the circuit.

Exercise 11.6

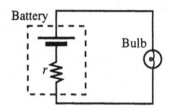

Consider a circuit in which a bulb is connected across a real battery. We will examine what happens to the circuit when wires of different resistance are used to short the battery.

For this exercise, assume the battery has a constant internal resistance of 0.01 ohm and an open-circuit voltage of 1.5 volts. We will pretend that the bulb has a constant resistance of 5.00 ohms.

A. Calculate the current and voltage for each element in the circuit above (when there is no short across the battery). Record your values in the first row of a table like the one below.

Wire resistance	Bulb		Wire		Internal resistance		Battery	
	voltage	current	voltage	current	voltage	current	voltage	current
no short								
0.1 Ω								
0.01 Ω								
0.001 Ω								

McDermott & P.E.G., U.Wash./*Physics by Inquiry*

B. Now imagine that the battery is shorted
by a wire as shown. Find the voltage
and current for each element when the
shorting wire has each of the following
resistances: 0.1 Ω, 0.01 Ω, and
0.001 Ω.

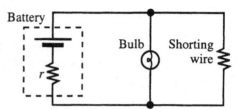

Record your results in your table.

C. Refer to your table to answer the following:

What happens to the current through a battery when it is shorted by
a wire with a resistance (1) larger than the internal resistance of the
battery? (2) about the same size as the internal resistance?
(3) smaller than the internal resistance?

✔ Discuss your results with a staff member.

Exercise 11.7

For the circuit at right, assume that
the resistance of the bulb and the
internal resistance of the battery are
constant. Also assume that the bulb
has much greater resistance than the
battery.

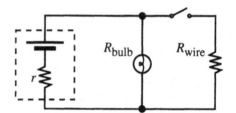

A. Assume that the resistance of the wire is small compared to the
resistance of the bulb, but large compared to the internal resistance of
the battery.

When the switch is closed, how does the voltage across the bulb
compare to the voltage across the load?

When the switch is closed, how does the resistance of the load
compare to the internal resistance of the battery? How does the
voltage across the load compare to the voltage across the internal
resistance of the battery?

Will the voltage across the bulb change significantly when the
switch is closed? Will the brightness of the bulb change
significantly?

B. Now assume that the resistance of the wire is about the same as, or less
than, the internal resistance of the battery.

When the switch is closed, how does the resistance of the load
compare to the internal resistance of the battery? How does the
voltage across the load compare to the voltage across the internal
resistance of the battery?

When the switch is closed, will the voltage across the bulb change significantly? Will the brightness of the bulb change significantly?

C. In this and the preceding exercise, we examined the effect on a circuit of connecting a wire across a battery.

Discuss the circumstances under which this results in a short circuit. Does a short result whenever a wire is connected across the battery?

Are there circumstances under which a connecting wire can be considered to be a short in one circuit and not in another circuit?

D. Discuss the following statement made by a student.

"A short circuit happens whenever a wire with small resistance is put in parallel with a bulb. The short draws current from the bulb, and so the bulb goes out."

✔ Discuss your reasoning with a staff member.

Exercise 11.8

The bulb in the circuit shown is shorted by a wire connected across the terminals of the battery. Two students offer explanations for the observation that the bulb goes out.

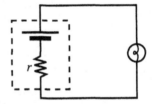

Student 1: *"Since the short has much less resistance than the bulb, most of the current will go through the wire so the bulb will not light."*

Student 2: *"Since the bulb goes out, the voltage drop across it must have been greatly reduced. This will happen if the resistance of the short is less than the internal resistance of the battery because then the resistance of the load will also be less than the internal resistance. Since the load and the internal resistance are in series, the voltages are in the same proportion as the resistances. Thus most of the voltage drop occurs across the internal resistance in the battery."*

Discuss the explanations given by student's 1 and 2. If either of the explanations are incomplete or incorrect, describe how you would help the student come to a correct explanation of why the bulb went out.

Exercise 11.9

During periods of high use of electricity, electrical systems sometimes experience a condition known as a brown-out. During a brown-out, the system voltage drops from 120 volts to, say, 105 volts.

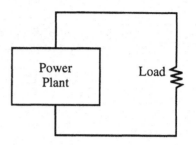

Consider the model of a municipal power system shown at right .

A. Does a brown-out occur when the load resistance R becomes large or small?

B. What might cause the resistance of the load to change in the way required to produce a brown-out?

C. Why does the voltage decrease when the load resistor changes as described in parts A and B?

Section 12. Energy and power

The light bulbs we have worked with thus far have all been identical. In real life there are
many different types of bulbs. In the following experiment we compare the behavior of
light bulbs that are not identical but have different filaments.

Experiment 12.1

Obtain two bulbs with different types of filaments.

A. Connect the two bulbs with different filaments in series across a
    battery. Observe and record their relative brightness.

    Use the model for current in an electric circuit to compare the
    currents through one the two bulbs.

    Is bulb brightness determined only by current? (i.e., Do all bulbs
    with the same current through them have the same brightness?)
    Explain.

B. Connect the two bulbs with different filaments in parallel across a
    battery. Observe and record their relative brightness.

    Compare the voltages across the bulbs.

    Is bulb brightness determined only by voltage? (i.e., Do all bulbs
    with the same voltage across them have the same brightness?)
    Explain.

C. Design and sketch a circuit to compare the resistances of the two bulbs
    with different filaments. (*Hint:* See the circuit used in Exercise 3.9 to
    compare the resistances of two "black boxes.") Do not use an
    ammeter.

    What is the total number of bulbs that you will need?

    Which elements in your circuit correspond to the "black boxes" in
    Exercise 3.9?

    Set up your circuit and compare the bulb resistances. Record your
    observations.

D. On the basis of your observations in parts A–C, answer the following
    question:

    Is bulb brightness determined only by the resistance of a bulb?
    Explain.

✔ Check your reasoning with a staff member.

McDermott & P.E.G., U.Wash./*Physics by Inquiry*

When two bulbs are identical, we have found that we can predict which will be brighter if we know which bulb has a greater current through it or which has a greater voltage across it. However, as the above experiment indicates, neither voltage alone nor current alone is sufficient to account for the brightness of bulbs that are not identical. In the series circuit, both bulbs have the same current; yet they have different brightness. In the parallel circuit, both bulbs have the same voltage across them; yet they are not equally bright. We also cannot explain the difference in brightness on the basis of resistance alone. Each bulb is brighter in one circuit and dimmer in the other. To explain the relative brightness of the bulbs in Experiment 12.1, a new concept is needed. In this section, we develop the tools necessary to predict the brightness of non-identical bulbs in an electric circuit.

Experiment 12.2

Set up the following four circuits using identical bulbs. Record the relative brightness of each of the bulbs.

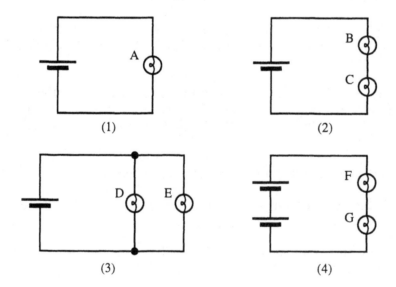

A. Compare the voltage across the load in each circuit. Explain.

B. Compare the current through the batteries in each circuit. Explain.

C. Leave the circuits connected for the entire class period. Check the circuits occasionally and make note of how the brightness of each bulb changes.

In which circuit does the battery last the shortest amount of time?

In which circuit does the battery last the longest amount of time?

Energy

Our model for electric circuits includes the idea that there is a flow in a continuous loop from the battery through the circuit and back through the battery. The current is not "used up" by the circuit elements. However, the warming of a wire connected across a battery; the production of light and heat by bulbs; and the running down of the battery are evidence that the battery is supplying the circuit with something that does not return to the battery. We call this quantity the *electrical potential energy,* or more briefly, the *electrical energy.*

Until now, we have focused on the role of the battery as the agent that drives current around the circuit. We can also consider the battery as the source of electrical energy for the circuit. Resistors and bulbs transform electrical energy into heat and light energy which are transferred to the surroundings.

A basic principle in physics is that energy is conserved. Although it may be changed from one form to another (for example, from chemical to electrical to heat), the total amount of energy remains the same. The equality between the electrical energy generated within the battery and the energy dissipated in the circuit is the basis of Kirchhoff's second rule. Chemical reactions within the battery release energy that is made available to light bulbs, and heat resistors. This process ceases when the substances within the battery can no longer sustain the chemical reactions and the battery "dies." It is in this sense that one may hear of the electrical energy as being "used up" by the elements in the circuit.

To a certain approximation, we can think of the battery as having a fixed amount of electrical energy that can be delivered to a circuit. (Although in actual practice not all of this energy is available to the circuit.) In the system of units we are using, the unit of energy is the *joule* (abbreviated J). In the following experiments we consider how we can use the idea of energy to predict the lifetime of batteries.

Exercise 12.3

Consider the two circuits shown below. Use the idea of energy to predict how the lifetimes of the batteries will compare. Explain your reasoning. Discuss how the relative brightness of the two bulbs is important in making your prediction.

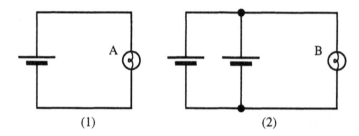

(1)   (2)

Exercise 12.4

In Experiment 12.2 we examined the lifetimes of the batteries in the circuits shown below.

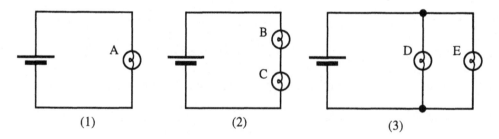

(1)   (2)   (3)

A. Compare the voltage across the load in the circuits above. Explain your reasoning.

Compare the current through the battery in the circuits above. Explain your reasoning.

B. Use the idea of energy to predict how the lifetimes of the batteries in circuit 1 and circuit 3 should compare. Discuss how your observation of the relative brightness of the bulbs is important in making your prediction.

> Are your observations of the lifetimes (see Experiment 12.2) consistent with your predictions?

> Does the lifetime of the battery appear to be related to the relative amount of current through the battery? Explain your reasoning.

C. A student makes the following comparison of the lifetimes of the batteries in circuit 1 and circuit 2.

> *"Since the bulbs in circuit 2 are identical, they each get half of the energy from the battery before it dies. When you add up the energy you get the same total energy as in circuit 1. So the batteries should die in the same amount of time."*

Discuss this statement with your partner. Do you agree with the reasoning used by this student?

✔ Check your reasoning with a staff member.

As the previous exercises illustrate, it is important to consider the rate at which energy is transferred from a battery when determining the lifetime of the battery. If a battery quickly transfers its energy to a circuit it will not last as long as if the energy is transferred more slowly. If one bulb is brighter than another, the brighter bulb is transferring energy more quickly to its environment; likewise, if one wire is warmer than another, the warmer wire is dissipating energy more quickly. In the following exercise we consider how the idea of the rate at which energy is transferred can be used to compare the lifetimes of batteries.

Exercise 12.5

In Experiment 12.2 you should have observed that the battery in the circuit
with two bulbs in series lasted about twice as long as the battery in the
circuit with a single bulb.

What does this observation suggest about the relative rates at which
energy is being transferred from the two batteries? Explain.

What does this observation suggest about the relative rates at which
the individual bulbs in the two circuits receive energy? Explain.

Power

The brightness of a bulb is closely tied to the rate at which it receives and emits the

energy that it receives from the battery. Usually the *rate* of energy use, the number of

joules of energy transferred in 1 second of time, is a more practical quantity than the total

amount of energy transferred. For a light bulb, the brightness is more important than the

total amount of light that is emitted from the time the bulb is turned on until it is turned

off. Similarly, the more important quantity describing a hydroelectric plant is the rate at

which it supplies energy, rather than the total energy it provides during its lifetime. The

energy rate is called the *power* and will be represented by the symbol *P*. The unit of

power is the joule per second, but this unit has its own name: the *watt* (W).

We interpret the power as the number of joules of energy that a battery transfers to the

circuit each second or that a circuit element transfers to the surroundings each second.

Usually we are more interested in the power delivered by the battery to the bulb or

resistor than in the total energy involved in the transfer. However, there are

circumstances in which we may want to know the total energy expended over a certain

period of time.

Exercise 12.6

Three students are discussing their electric bill.

> Student 1: *"Kilowatt-hours... That's the current we drew for the whole month."*
>
> Student 2: *"No. The kilowatt-hours is just the amount of electricity we used."*
>
> Student 3: *"Actually, that tells us the power we used."*

Which, if any, of these students is right? If none is exactly right, give a correct interpretation of the number of kilowatt-hours and explain why this is the quantity of interest in the calculation of an electric bill.

Exercise 12.4 illustrates that if the *voltage* across two loads is the same, then the power is proportional to the current through the battery. In some cases, however, we need to compare the power delivered by the battery in circuits where the *current* through the battery is the same, but the voltage across the load is different.

Exercise 12.7

Consider the following circuits from Experiment 12.2.

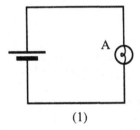
(1)

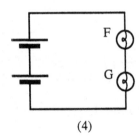
(4)

A. Compare the currents through the batteries. Explain.

Compare the voltage across the load in each circuit. Explain.

Compare the brightness of the three bulbs.

What does the relative brightness of the bulbs suggest about the rate at which energy is dissipated in circuit 4 compared to the power dissipated in circuit 1?

B. Suppose that circuit 4 now had three batteries in series with three bulbs in series. Repeat part A for this case.

          McDermott & P.E.G., U.Wash./*Physics by Inquiry*

Your observations in this section are consistent with the idea that the rate at which electrical energy is dissipated in a light bulb (or a resistor) depends on both the current through it and on the voltage across it. We found that if the voltage across a load is fixed, the power dissipated is proportional to the current through the load; if the current is fixed, the power is proportional to the voltage across the load. For any load, we write this as $P_{\text{load}} = V_{\text{load}} i_{\text{load}}$, where V_{load} is the voltage across the load and i_{load} is the current in it.

Exercise 12.8

A. There is a current i through a resistor of resistance R. Write an expression in terms of i and R for the power delivered to the resistor.

B. A voltage V is maintained across a resistor of resistance R. Write an expression in terms of V and R for the power delivered to the resistor.

Now we are in a position to account for our observations of the brightness of the two non-identical bulbs in Experiment 12.1.

Exercise 12.9

On the basis of the model we have developed, account for the relative brightness of the two non-identical bulbs in Experiment 12.1:

• when the two bulbs are connected in series with a battery

• when the two bulbs are connected in parallel with a battery

✔ Explain your reasoning to a staff member.

Exercise 12.10

In the circuits below, assume all the batteries and resistors are identical.
Rank the circuits according to:

- total energy output of the circuit until the battery or batteries die

- lifetime of the batteries

- total power delivered to the resistors in the circuit

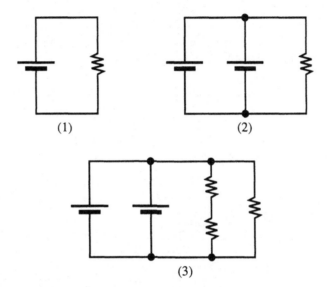

(1)   (2)

(3)

✔ Explain your reasoning to a staff member.

Supplementary problems for *Electric Circuits*

Note: In all problems assume the bulbs are identical and the batteries are identical unless otherwise indicated.

Problem 1.1

For current to flow in an electric circuit, there must be a complete path of conductors from one terminal of the battery to the other. Examine a bulb and describe the path of conductors through the bulb. Make a detailed diagram showing this path.

Problem 1.2

Describe in words how a light socket works. Include a diagram showing the path of conductors in a socket that allows the socket to work.

Problem 1.3

Draw the circuit diagram for each of the following circuits.

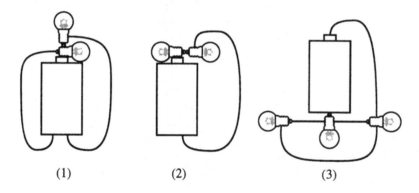

(1)   (2)   (3)

Problem 1.4

Draw a sketch of a bulb. On your sketch, indicate the two terminals of the bulb.

The symbol for a bulb introduced in this text (as shown at right) shows no difference between the two terminals. However, in the sketch you are making, the two terminals of the bulb should look quite different. Explain why we are justified in using a symbol for a bulb in which the two terminals are drawn in the same way.

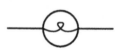

Problem 2.1

Rank the following circuits in order by the current through the battery. Explain your reasoning.

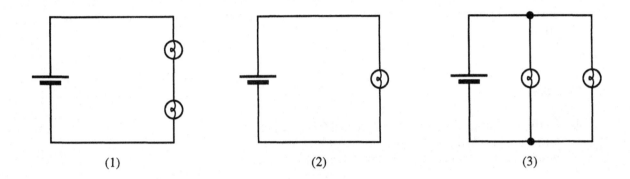

    (1)   (2)   (3)

Problem 2.2

Consider the following dispute between two students. Which student is correct?

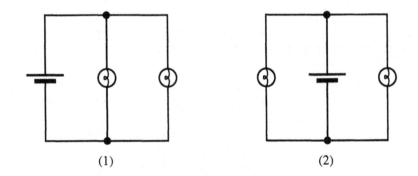

    (1)   (2)

Student 1:   *"Circuit 1 and circuit 2 are different circuits. In circuit 2 the bulbs are the same distance from the battery, but in circuit 1 one bulb is closer to the battery. So the brightness of the bulbs could be different in the two circuits."*

Student 2:   *"Circuit diagrams don't show physical layout, only electrical connections. In each diagram, the electrical connections are the same. Each bulb has one terminal directly connected to one terminal of the battery and the other terminal of each bulb is directly connected to the other terminal of the battery. So the brightness of the bulbs will be the same in each circuit. Both diagrams represent the same circuit."*

Problem 2.3

Use the results of your experiments and the assumptions we have made for electric circuits to justify the following statement:

> "The current through a battery isn't constant but depends on the circuit. The current can be less than or more than the current through a bulb in a single-bulb circuit."

Problem 3.1

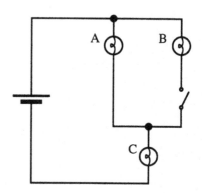

A. In the circuit at right, use your model for electric current to rank the bulbs according to brightness when the switch is open. Explain your reasoning.

B. Rank the bulbs according to brightness when the switch is closed. Explain your reasoning.

Problem 3.2

Below is a diagram showing a typical household circuit. The appliances (lights, television, toaster, etc.) are represented by boxes labeled 1, 2, 3, and so on.

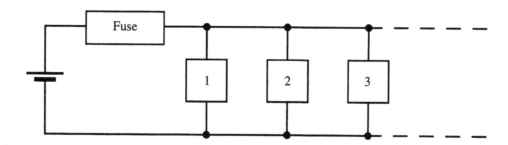

A fuse or circuit breaker is intended to shut off the circuit if the wires in the circuit get hot enough to pose a threat of fire. The wires in a circuit (which in a house are in the walls) are hotter when more current is flowing through them. A fuse or circuit breaker is sensitive to the amount of current passing through it. A fuse will heat up and burn out (melt) when too large a current flows through it. A circuit breaker will open a switch and break the circuit when too large a current flows.

A. What happens to the current through the fuse in the circuit above when more appliances are added to the circuit?

B. What may happen to the fuse if too many appliances are added to the circuit?

Problem 3.3

In the circuit below, use your model for electric current to rank the bulbs according to brightness with the switches in all possible combinations of positions. Explain your reasoning.

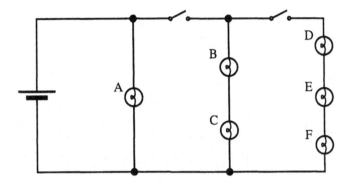

Problem 3.4

A. Consider the circuit diagram below. What would happen to the brightness of the other bulbs in the circuit if one of the bulbs were to burn out (so that no current could flow through it)? Explain your reasoning.

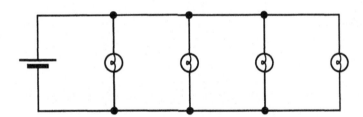

B. Consider the circuit diagram below. Again, what would happen to the brightness of the other bulbs in the circuit if one of the bulbs were to burn out?

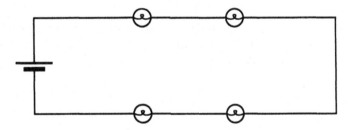

C. Which arrangement of bulbs do you think would work better for a string of ornamental lights? Explain your reasoning.

     McDermott & P.E.G., U.Wash./ *Physics by Inquiry*

Problem 3.5

A. Identical bulbs are connected to identical batteries in the various circuits below. Use your model and previous observations to rank all the bulbs (A–H) in order by brightness. If two bulbs have the same brightness, state that explicitly. Explain your reasoning.

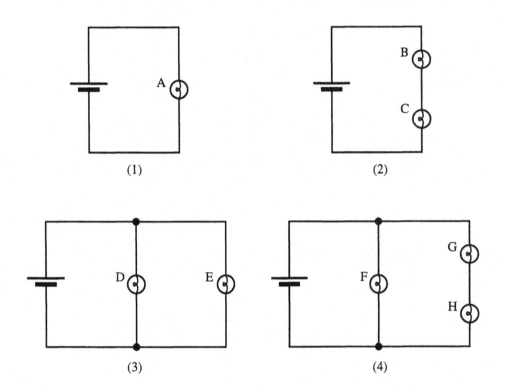

B. Rank the bulbs according to the current through each bulb. Explain your reasoning.

C. Rank the circuits according to the current through the battery. Explain your reasoning.

D. Rank the circuits according to the obstacle to the flow of current presented by the various combinations of bulbs, that is, rank them in order of the *total* resistance in each circuit. Explain your reasoning.

Problem 4.1

State whether the bulbs in the following circuits are arranged in series, parallel, or neither, for each possible combination of switch positions.

Predict how the brightness of bulbs A and B would compare in each case and explain your reasoning.

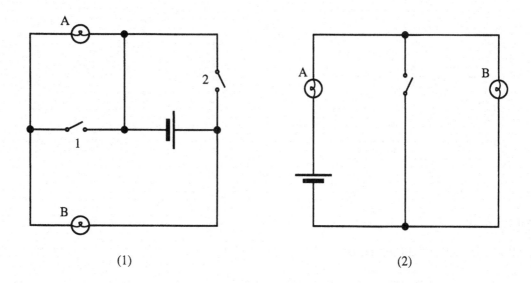

(1)   (2)

Problem 4.2

State whether the bulbs in the following circuits are arranged in series, parallel, or neither, for each possible combination of switch positions.

Predict how the brightness of bulbs A and B would compare in each case and explain your reasoning.

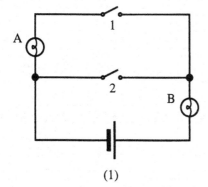   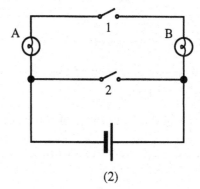

(1)   (2)

    McDermott & P.E.G., U.Wash./ *Physics by Inquiry*

Problem 4.3

State whether the bulbs in the following circuits are arranged in series, parallel, or neither, for each possible combination of switch positions.

Predict how the brightness of the bulbs would compare in each case and explain your reasoning.

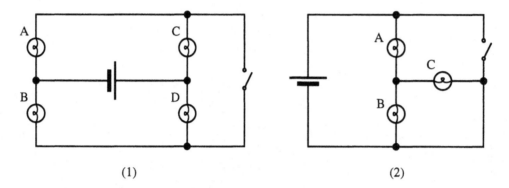

(1)   (2)

Problem 4.4

Consider the circuit shown below.

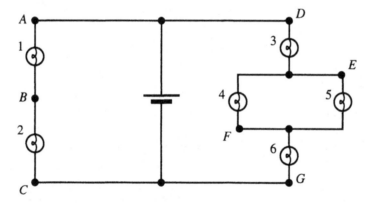

A. Rank the bulbs according to brightness. Explain your reasoning.

   After answering each of the following parts, return to the original circuit before going to the next part.

B. How will the brightness of bulbs 1 and 3 *change* if we unscrew bulb 4? Explain.

C. How will the brightness of bulbs 2, 5, and 6 *change* if we add a new bulb by connecting it between points *D* and *E?* Explain.

D. How will the brightness of bulbs 1, 3, 5, and 6 *change* if we connect a conductor between points *A* and *F?* Explain.

E. How will the brightness of bulbs 1 and 6 *change* if we add a new bulb by connecting it from point *C* to point *G* (without changing any of the connections shown)? Explain.

Problem 4.5

Below are two circuit diagrams drawn in an unorthodox way.

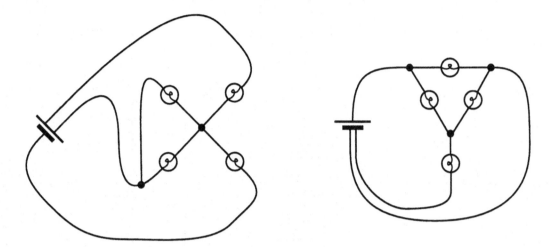

Pick the standard circuit diagrams below that represent the circuits above.

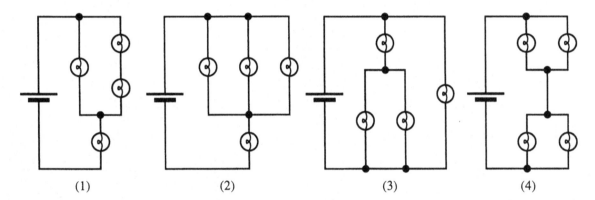

(1)   (2)   (3)   (4)

Problem 4.6

State whether the bulbs in the following circuits are arranged in series, parallel, or neither, for each possible combination of switch positions.

Predict how the brightness of the bulbs would compare in each case and explain your reasoning.

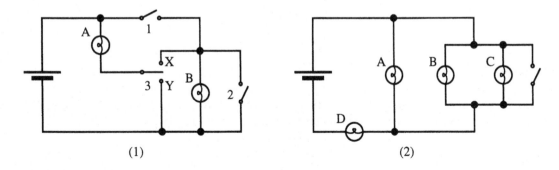

(1)   (2)

McDermott & P.E.G., U.Wash./ *Physics by Inquiry*

Problem 4.7

State whether the bulbs in the following circuits are arranged in series, parallel, or neither, for each possible combination of switch positions.

Predict how the brightness of the bulbs would compare in each case and explain your reasoning.

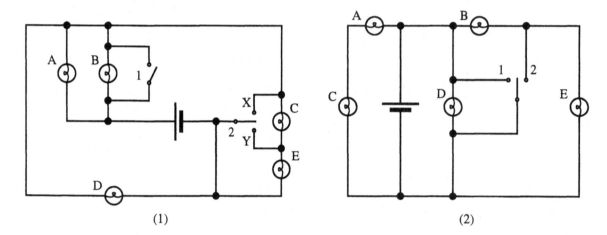

(1)   (2)

Problem 5.1

Consider the following two circuits.

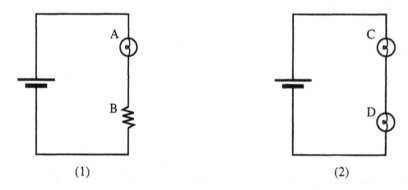

(1)   (2)

Circuit 1 contains a bulb and a length of nichrome wire. Circuit 2 contains 2 bulbs, identical to the bulb in circuit 1. Bulbs A and C have the same brightness. What can you conclude about how much resistance the bulbs in these circuits offer to the flow of electric current compared to the resistance offered by the nichrome wire? Explain your reasoning.

Problem 5.2

A total current i divides into two branches, A and B. Branch A carries 2/3 the total current. What is the ratio of the current in branch A compared with the current in branch B?

Problem 5.3

Consider the four-way node shown on the right. The total current flowing into the node is *i*. The current in branch A is three times the current in branch B. The current in branch C is two times the current in branch B. Write algebraic expressions in terms of *i* for the current in each branch.

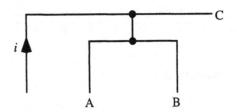

Problem 5.4

What is the current through each of bulbs in the circuit shown on the right? Explain your reasoning.

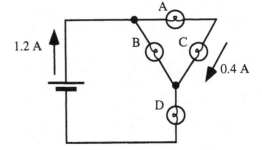

Problem 5.5

A. In the network above, rank the bulbs according to brightness. Explain your reasoning. (Note that you may not be able to compare all the bulbs with each other. Do the ranking only for the bulbs you *can* compare.)

B. How will the brightness of bulbs 1, 3, and 7 *change* if we add a length of nichrome wire in series with bulb 6? Explain your reasoning.

C. How will the brightness of bulbs 1, 4, and 7 *change* if we add a length of nichrome wire by connecting it between points *D* and *H?* Explain your reasoning.

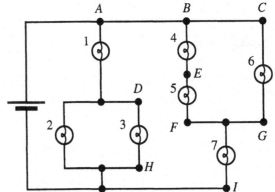

D. How will the brightness of bulbs 1, 2, 6, and 7 *change* if we connect a conductor between points *B* and *H?* Explain your reasoning.

Problem 5.6

The currents flowing out of a four-way node are *i* and 3*i*/4. The currents flowing into the node along the two other paths are in the ratio of 2 to 5. Write expressions in terms of *i* for the currents flowing into the node.

McDermott & P.E.G., U.Wash./ *Physics by Inquiry*

Problem 5.7

A total current i divides into two branches. The currents in the two branches are in the ratio of x to $x+1$. Write expressions in terms of i and x for the current in each branch.

Problem 6.1

Determine the equivalent resistance of each of the following networks. Explain your reasoning.

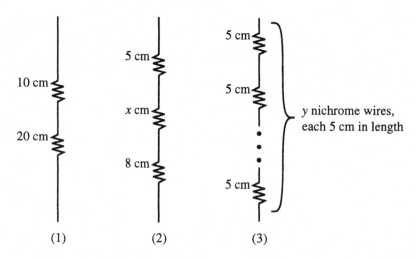

Problem 6.2

Determine the equivalent resistance of each of the following networks. Explain your reasoning.

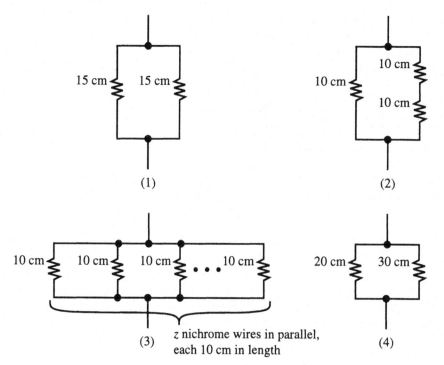

Problem 6.3

Suppose you have a circuit element, Q, that has a resistance that is not constant. In particular, suppose that as the current through this element increases, the resistance of the element also increases.

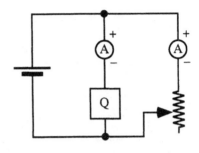

Your partner sets up the circuit shown at right. Your partner finds that a length L of nichrome wire results in the same current through both branches.

A. Imagine that your partner replaces element Q by two such elements connected in series, as shown at right. Your partner then finds the length, L', of nichrome wire that results in the same current through both branches.

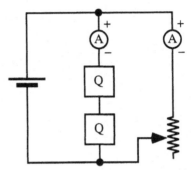

> Would the length L' be equal to $2L$, greater than $2L$, or less than $2L$? Explain your reasoning.

B. Now imagine that your partner replaces the series network with a parallel network consisting of the same two elements, as shown at right. Your partner finds that the length of nichrome wire that results in the same current through both branches is L''.

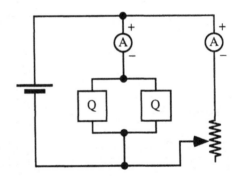

> Would the length L'' be equal to $L/2$, greater than $L/2$, or less than $L/2$? Explain your reasoning.

Problem 7.1

For the circuit at right, compare the voltages across BC, AD, AE, BF, and AC. What is the voltage AB? CE? Explain your reasoning.

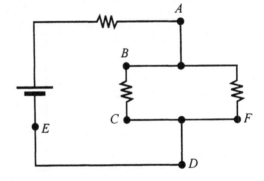

Problem 7.2

A network consisting of two resistors in series has a total voltage V across it. The voltage across the first resistor is one-half as great as the voltage across the second. Write expressions in terms of V for the voltage across each resistor.

     McDermott & P.E.G., U.Wash./ *Physics by Inquiry*

Problem 7.3

A. Consider circuit 2 and circuit 3 below.

How does the voltage across the top bulb in each circuit compare to the voltage across the bottom bulb in each circuit?

How does the current through the top bulb in each circuit compare to the current through the bottom bulb in each circuit?

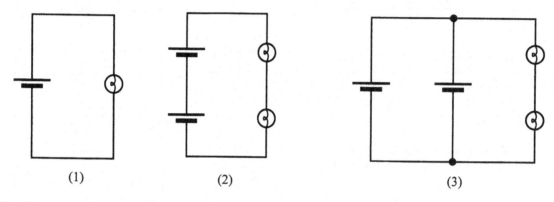

(1)   (2)   (3)

B. Compare the current through the bulbs for the three circuits shown.

In which circuit is the current through the bulbs the greatest? the least?

In which circuit is the voltage across the pair of bulbs the greatest? the least?

Problem 7.4

For the circuits shown below, compare the brightness of bulbs X, Y, and Z. Compare the current through batteries A, B, and C. Explain your reasoning.

How does the voltage across bulb X compare to the voltage across bulbs Y and Z? Explain your reasoning.

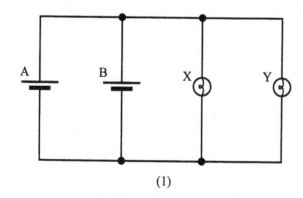   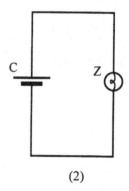

(1)   (2)

Problem 8.1

What is the voltage across each of the lettered elements in the following circuits?

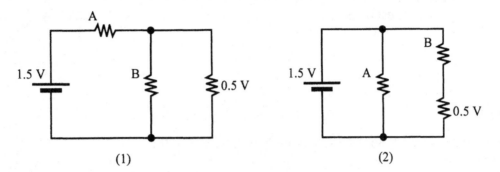

(1)   (2)

Problem 8.2

Determine the voltage across each of the lettered elements in the following circuits. Explain your reasoning.

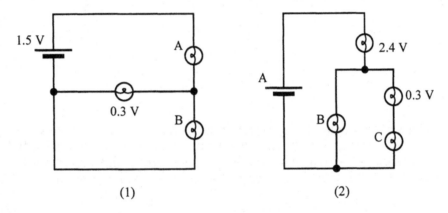

(1)   (2)

Problem 8.3

A. Rank the bulbs in the circuit below according to brightness. Explain your reasoning.

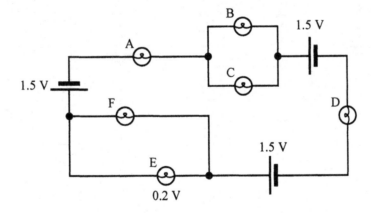

B. Using Kirchhoff's second rule, find the voltage across each of the bulbs. Explain your reasoning.

McDermott & P.E.G., U.Wash./ *Physics by Inquiry*

Problem 8.4

In the circuit at right, the voltage across resistor 1 is x times the voltage across resistor 2. Write an expression in terms of x and V for the voltage across each resistor.

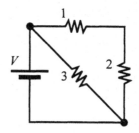
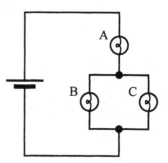

Problem 8.5

Suppose that another bulb is added in parallel to bulb B in the circuit at right:

A. Predict how the voltages across bulbs A, B, and C would change.

B. Predict how the currents through bulbs A, B, and C would change.

Explain your reasoning.

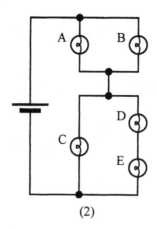

Problem 8.6

Predict the relative brightness of the bulbs within each of the circuits below. (Do not compare the bulbs in circuit 1 with the bulbs in circuit 2.) Attempt to decide bulb brightness in each circuit by reasoning on the basis of current and also by reasoning on the basis of voltage. Which method is better suited for analyzing each circuit? Explain.

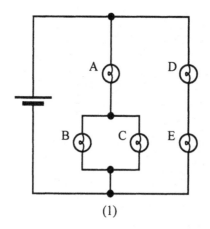
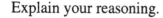

(1)   (2)

McDermott & P.E.G., U.Wash./ *Physics by Inquiry*

Problem 9.1

Decompose the circuit at right into its series and parallel networks.

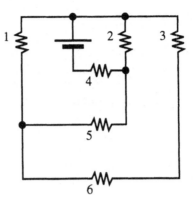

A. What happens to the current in resistor 3 if resistor 1 is replaced by a conductor? Explain.

B. What happens to the current in resistor 3 if resistor 2 is replaced by a conductor? Explain.

Problem 9.2

Decompose the circuit at right into its series and parallel networks.

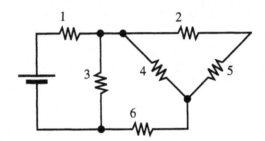

What happens to the current in resistors 1 and 6 if the resistance of resistor 5 is increased? Explain your reasoning.

Problem 9.3

Decompose the circuit at right into its series and parallel networks.

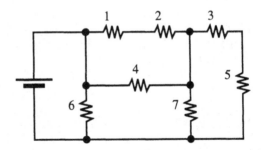

What happens to the current in resistors 1, 3, 4, and 6 if resistor 5 is replaced by a connecting wire? Explain your reasoning.

Problem 9.4

Decompose the circuit at right into its series and parallel networks.

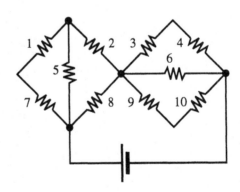

Predict what would happen to the current in resistor 2 if a bulb were added in parallel to resistor 7.

McDermott & P.E.G., U.Wash./ *Physics by Inquiry*

Problem 9.5

A. Decompose the circuit at right into its series and parallel networks. Show clearly how you made your decomposition.

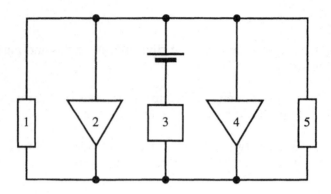

B. Suppose network 3 consists of the arrangement of bulbs shown below as X and network 5 consists of the arrangement Y. How will the current through bulb A change if network 5 is replaced by arrangement Z? Explain your reasoning.

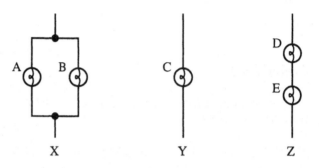

Problem 9.6

Decompose the circuit at right into its series and parallel networks.

Predict what would happen to the current in resistor 4 if a bulb were added in parallel to resistor 6.

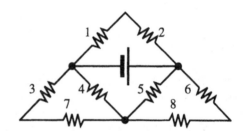

Problem 9.7

Decompose the following circuit into its series and parallel networks.

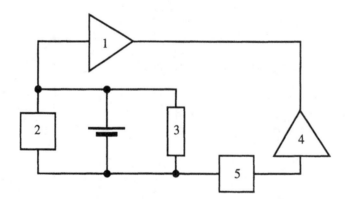

Describe what would happen to the current through network 2 if the resistance of network 4 were increased. Explain your reasoning.

Problem 9.8

Determine the current through each of the non-identical bulbs in the circuit shown below. Explain your reasoning.

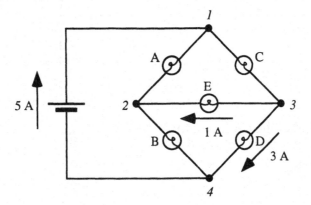

Problem 9.9

The current through resistor C is 0.6 A. The current through resistor A is twice as large as the current through resistor D. The current through resistor B is one-third as big as the current through resistor E.

Find the current through the battery. Explain your reasoning. (This includes clearly defining in words all your variables and saying in words why two expressions can be set equal.)

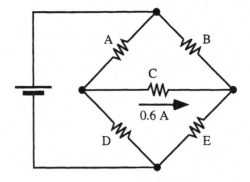

Problem 10.1

Rank the following networks in order of resistance. Assume all the linear resistors are identical. Explain your reasoning carefully.

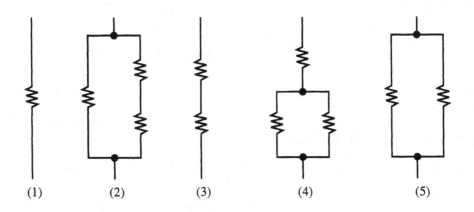

    McDermott & P.E.G., U.Wash./ *Physics by Inquiry*

Problem 10.2

In the circuit at right, find the current through each element and the voltage across each element. Explain your reasoning.

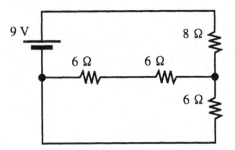

Problem 10.3

In the circuit at right, find the current through each element and the voltage across each element. Explain your reasoning.

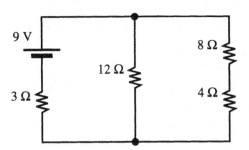

Problem 10.4

In the circuit at right, find the current through each element and the voltage across each element. Explain your reasoning.

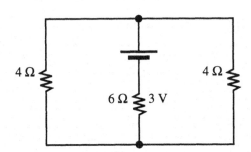

Problem 10.5

A. In the circuit at right, find the voltage across nodes *A* and *B*.

B. Find the current through the 8-ohm resistor connected between nodes *A* and *C*.

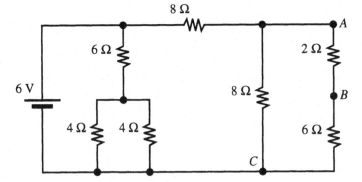

Problem 10.6

In the circuit at right, find the current through each element and the voltage across each element. Explain your reasoning.

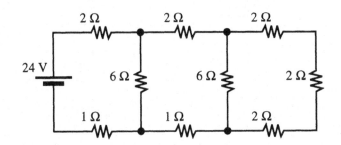

McDermott & P.E.G., U.Wash./ *Physics by Inquiry*

Problem 10.7

A. Find the voltage across each resistor in the circuit at right without calculating any currents. Explain your reasoning.

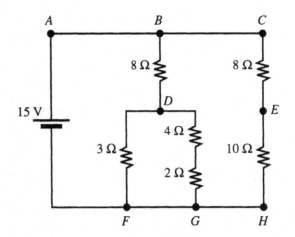

B. Find the current through each resistor in the following circuit without calculating any voltages. Explain your reasoning.

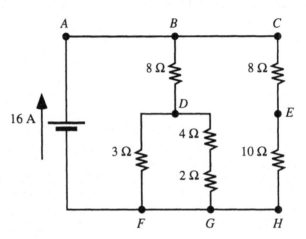

Problem 10.8

In the circuit at right, resistor A has 2.5 times as much current through it as resistor C and one-fourth as much voltage across it as resistor C.

A. Find the voltage across resistor B.

B Find the current through resistor B.

C. Find the resistance of resistor B.

Express your answers in terms of V and R_C.

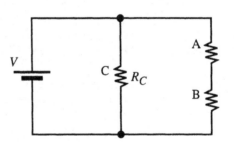

Problem 10.9

A. In the circuit at right, rank the resistors according to the amount of current through them. Explain your reasoning.

B. What happens to the current through each of the other resistors if the resistance of resistor C is increased? Explain your reasoning.

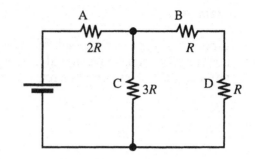

Problem 10.10

In the circuit at right, find the current through each element and the voltage across each element. Explain your reasoning.

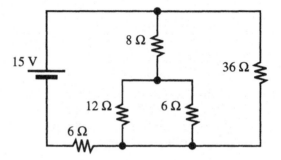

Problem 11.1

Consider a circuit in which a bulb is connected across a real battery. The battery has a constant internal resistance of $0.1\ \Omega$ and an open-circuit voltage of 1.5 V. Assume that the bulb has a constant resistance of $5.0\ \Omega$ and that the bulb is visibly lit only if the current through it is greater than 0.1 A.

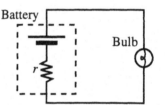

A. Find the current through the bulb. Is the bulb lit? Explain your reasoning.

B. How many identical bulbs can be connected in parallel with the original bulb before the bulb goes out? Explain your reasoning.

C. Now imagine that the battery is shorted by a wire as shown. Find the resistance of the wire for which the bulb will barely light.

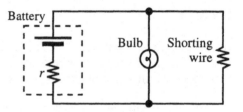

D. Suppose that the resistance of the shorting wire in Part C were increased. Would the brightness of the bulb increase, decrease, or remain the same? Explain your reasoning.

Problem 11.2

Suppose that you had at your disposal only a real battery, a voltmeter, an ammeter, a wire-cutter, and a piece of wire of unknown resistance.

Devise a method to determine the internal resistance of the battery. Explain your reasoning.

                    McDermott & P.E.G., U.Wash./ *Physics by Inquiry*

Problem 11.3

Two batteries A and B have the same open-circuit voltage of 1.5 volts but different internal resistances, $r_A = 0.1\ \Omega$ and $r_B = 0.4\ \Omega$. The batteries are connected in series with a resistor with resistance R. Find the value of R such that battery B has zero terminal voltage. Explain.

Problem 12.1

Assume the batteries shown below are identical and have no internal resistance. Each battery is capable of delivering 7200 J of energy to its circuit before it dies.

A. Compare the lifetimes of the two batteries. Explain your reasoning.

B. How long will each battery last? Explain your reasoning.

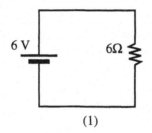

(1)

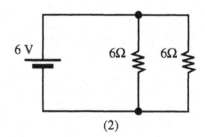

(2)

Problem 12.2

In the circuit shown, 189 J of energy are received by the resistor with resistance R in a time of 1 minute and 3 seconds when the current in the circuit is 1.5 A.

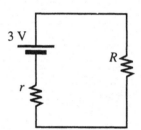

A. What is the voltage across the resistor with resistance R? Explain your reasoning.

B. How much energy is delivered each second to the resistor with resistance r? Explain your reasoning.

C. Find the resistance of each of the two resistors. Explain your reasoning.

Problem 12.3

In the circuit shown, a resistor with resistance R is connected to a 3 V battery with internal resistance $0.2\ \Omega$.

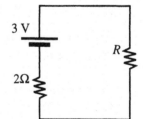

A. What is the power delivered to the resistor if:

- R is less than the internal resistance of the battery (e.g., $0.1\ \Omega$)? Explain your reasoning.

- R is larger than the internal resistance of the battery (e.g., $0.4\ \Omega$)?

- R is equal to the internal resistance of the battery?

B. To make the battery die as fast as possible, which of the above values of R would you choose? Explain your reasoning.

  McDermott & P.E.G., U.Wash./ *Physics by Inquiry*

Problem 12.4

A. Rank the resistors in the circuit shown according to:

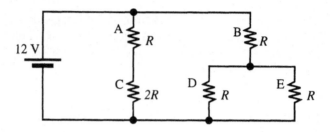

(1)   the current through them

(2)   the voltage across them

(3)   the power delivered to them

Explain your reasoning in each case.

B. Does the power dissipated in each of the other resistors increase, decrease, or remain the same when the following changes are made:

(1)   resistor E is shorted

(2)   the resistance of resistor C is increased

(3)   the resistance of resistor D is decreased

Explain your reasoning in each case.

Problem 12.5

In the circuit shown, resistor C has a resistance R and the voltage across the battery is V. The power delivered to resistor C is 3 times as great as the power delivered to resistor A. The current in resistor C is 0.25 times as great as the current in resistor B.

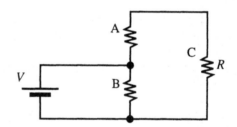

A. Write an expression in terms of V for the voltage across resistor B. Explain your reasoning.

B. Write an expression in terms of V and R for the current through resistor A. Explain your reasoning.

C. Write expressions in terms of R for the resistance of each of the resistors A and B. Explain your reasoning.

Problem 12.6

The circuits in this problem use two regular household light bulbs. One of the bulbs is a 60 watt bulb and the other is a 100 watt bulb. The power rating of household bulbs are this value when plugged into a 120 V wall outlet.

A. Which bulb will be brighter if the two are connected in parallel? Explain.

B. Which bulb will be brighter if the two are connected in series? Explain.

Problem 12.8

Explain your reasoning in answering the following questions about the circuit shown at right. How will the power delivered to resistor 3 change if:

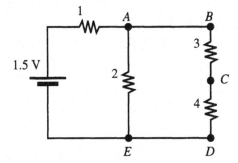

A. the resistance of resistor 2 is decreased?

B. an additional connecting wire is connected between points A and E?

C. a 10 ohm resistor is added in series with resistor 1?

D. a 10 ohm resistor is added in parallel with resistor 2?

PHYSICS BY INQUIRY

Electromagnets

In *Electromagnets*, we extend the study of magnetic interactions that we began in *Magnets* in Volume I. We examine the magnetic field around a current-carrying wire and make an electromagnet (a magnet that depends upon the presence of an electric current). We apply the findings from our investigation in the construction of a current meter and an electric motor.

Note: *Electromagnets* builds directly on *Magnets* in Volume I and on Parts A and B of *Electric Circuits* in Volume II. Students should finish working through these materials before beginning *Electromagnets*.

Section 1. Magnetic field of a current-carrying wire

Experiment 1.1

Use a ring stand and clamp to hold a piece of cardboard horizontally as
shown. Thread connecting wire through a hole in the cardboard, then
connect the wire to a battery and switch. Do not close the switch yet.

 Caution: Whenever you close the switch, the wire and battery
may become hot. Keep the switch open for all but the brief
periods during which you are making observations.

Place several small compasses on the cardboard around the wire.

A. Draw a sketch to show how the compasses are oriented before the
switch is closed. If the north poles of all the compasses are not
aligned, determine why not and adjust your setup accordingly.

B. Hold the wire vertically as shown. Close the switch *briefly* and
observe the behavior of the compass needles. After you open the
switch, record your observations.

We say that there is a magnetic field due to a current-carrying wire.
Explain how you can tell.

How does the strength of the magnetic field due to the current-
carrying wire vary with the distance from the wire? Explain how
you can tell.

McDermott & P.E.G., U.Wash./*Physics by Inquiry*

C. Make a sketch that shows magnetic field lines near the current-carrying wire.

Discuss:

- how you decided to draw the lines as you did

- how the field lines that you drew differ from those of a bar magnet

- how you took into account the magnetic field of the earth

(If necessary, review the similar experiments with bar magnets that appear in *Magnets* in Volume I.)

Compare the magnetic field for a current-carrying wire to the magnetic fields of other objects, such as a horseshoe magnet or a pair of bar magnets.

Are any of the magnetic fields that you have previously studied similar to the field of a current-carrying wire?

D. Reverse the connecting leads to the battery and repeat part B. Record your observations.

Sketch the magnetic field due to the current-carrying wire in this case.

How does reversing the leads to the battery affect the magnetic field due to the current-carrying wire?

Experiment 1.2

A student makes the following statement about the magnetic field of a current-carrying wire:

"The compasses are lying in a plane perpendicular to the wire so the compasses show the magnetic field lines as circles around the wire, but really the field lines of the current-carrying wire look like spirals winding up around the wire."

Design and perform an experiment to prove or disprove this student's statement.

✔ Check your work with a staff member.

Exercise 1.3

A. In *Electric Circuits*, we found that reversing the orientation of the battery does not affect the behavior of the battery, bulbs, and wires in a circuit.

Discuss how the results from Experiment 1.1 provide a method for distinguishing between the two battery orientations.

Although we can distinguish between the two orientations of a battery in a circuit, we have no way of knowing the direction of the electric current through the circuit. The standard convention is to imagine a flow from the positive terminal of the battery, through the circuit and back through the battery to the positive terminal.

B. Formulate a rule that will enable you to predict the direction of the magnetic field at any point near a straight current-carrying wire. Base your rule on the observations you have made thus far in this section.

✔ Check your work with a staff member.

Section 2. Making magnets with a current-carrying wire

In Section 1, we investigated the magnetic field of a straight current-carrying wire. In

this section, we examine how the magnetic field is affected when the shape of the wire is

changed.

Experiment 2.1

Wrap a 20 cm piece of insulated wire around a pencil several times.
Remove the pencil and place the wire coil on the cardboard from
Experiment 1.1. Connect the coil to a fresh battery through two holes in
the cardboard. Do not close the switch yet.

Place several compasses around the coil as shown.

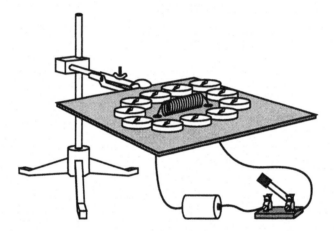

A. Before you close the switch, sketch the coil and compasses in your
notebook. Draw arrows to show the orientations of the compass
needles when the switch is open.

B. Briefly close the switch. Quickly observe the orientations of the
compass needles, then open the switch.

Make a sketch of the coil. Indicate the direction of the current through
each winding of the coil. Sketch small arrows to show the orientations
of the compass needles around the coil.

Sketch field lines to represent the magnetic field due to the current-
carrying coil. Base your sketch on your observations of the
orientations of the compass needles.

How do these field lines compare to those of a bar magnet?

Can you identify a north and south pole of the current-carrying
coil? Explain your reasoning.

C. Suppose you wanted to replace the current-carrying coil with a magnet that produces a similar magnetic field.

Describe the approximate location of the poles, the size, and the orientation of the magnet.

D. Reverse the leads to the battery in the circuit and repeat part B.

Describe how the magnetic field lines in this case differ from those you sketched in part B.

Experiment 2.2

In this experiment, you will investigate how the magnetic field of a current-carrying coil is affected by certain changes.

A. Connect a fresh battery, switch, and wire coil as shown. Determine how the strength and direction of the magnetic field near the ends of the coil are affected by the following changes. In each case, only close the switch briefly. Record your observations.

• Add various circuit elements to the circuit (e.g., a battery or a bulb).

• Change the shape, size, or number of turns of the wire coil.

• Put various materials inside the coil (e.g., insert a pencil or a nail).

B. Which change(s) above had a significant effect on the strength of the magnetic field of the current-carrying coil? Explain how you can tell.

Experiment 2.3

Wrap a piece of insulated wire around a ferromagnetic object (e.g., a nail) that is not a permanent magnet. Connect the wire to a battery and switch as shown. Use a fresh battery for this experiment.

Do not close the switch yet.

A. While the switch is open, bring the nail near several paper clips. If the nail attracts paper clips, find a nail that is not a magnet.

B. Bring the nail near some paper clips and then briefly close the switch. Describe your observations.

>While the switch is closed, does the nail attract paper clips? If so, how many paper clips can the nail pick up?

>Are the paper clips still attracted to the nail after the switch is opened? If so, does the nail pick up as many as when the switch is closed?

C. Predict the locations of the north and south poles of the nail-coil combination while the switch is closed. (*Hint:* Base your prediction on your observations from Experiment 2.1)

Obtain a permanent magnet and check your prediction.

D. Use your model for magnetic materials to account for the behavior of the nail (1) while the switch is closed and (2) after the switch has been opened. (*Hint:* Recall how magnets can be made from ferromagnetic objects. For examples of this process, see Section 8 of *Magnets* in Volume I.)

✔ Check your reasoning with a staff member.

Experiment 2.4

A. Make a coil as shown by winding many turns of insulated wire. The loops should form a short stack rather than be spread out in a spiral as in Experiment 2.1. Connect the coil to a battery and switch. Do not close the switch yet.

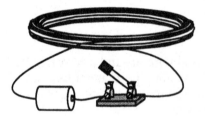

Sketch the coil and circuit in your notebook. On the diagram, indicate the direction of the current through the coil when the switch is closed.

B. Suppose a magnet were suspended over the coil as shown. This arrangement would allow the magnet to rotate about a horizontal pivot.

Predict whether the magnet will rotate when the switch is closed. If you believe the magnet will rotate, predict the direction in which it will rotate. Explain the basis for your prediction.

Suspend the magnet above the coil and check your prediction.

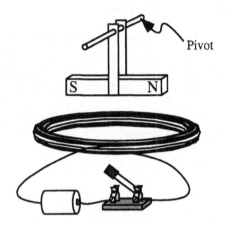

C. Predict what will happen when the orientation of the battery is reversed, then check your prediction.

Predict what will happen if a second battery is added in series with the first, then check your prediction.

Predict whether doubling the number of windings of the coil will affect the strength of the magnetic field of the coil. Explain your prediction and check your answer.

D. Can you identify a north and south pole for the coil when the switch is closed? How are the poles affected by the orientation of the battery? Explain.

If you were to replace the current-carrying coil by a magnet that produces a similar magnetic field, describe the location of the poles, the size and the orientation of the magnet.

E. Hold the coil so that the axis of the coil is oriented horizontally.

Use a small compass to determine the direction of the magnetic field at the center of the coil while the switch is closed.

(1) How does your result compare to the direction of the field along the axis of the coil, but outside the windings?

(2) What do your observations suggest about the direction of the magnetic field inside a magnet?

(3) In an experiment in *Magnets*, one magnet is removed from the center of a stack of small magnets. A compass is placed at the location of the missing magnet. From the behavior of the compass, the direction of the field inside a bar magnet is inferred.

Are the results from this experiment consistent with your answer in part (2) above?

F. Obtain a small box and lay the wire coil flat on the bottom of the box. Push two brads through the front of the box as shown. Attach the wire leads of the coil to the two brads, and connect the battery and switch to the brads as shown. Cut notches in the top of the box as shown to hold the dowel that supports your permanent magnet.

Describe how this device can be used to indicate the relative magnitude and direction of current in a circuit.

How is it similar to an ammeter? How does it differ?

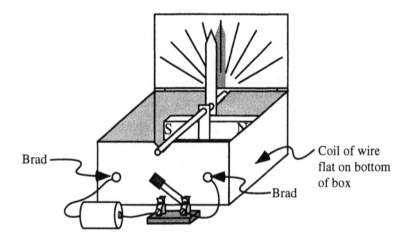

Section 3. Building motors

Experiment 3.1

A. Wind a long piece of sturdy insulated
wire around a soda can several times and
then remove the can so that you have a
wire coil with several loops. Wrap the
ends of the wire around the loops so that
they do not separate. Leave enough wire
at the ends of the coil to be able to bend
them into support leads as shown.
Remove the insulating material from the leads of the coil.

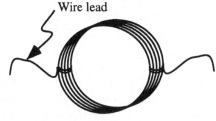

B. Obtain a metal wire clothes hanger and cut it to form conducting
supports for the coil as shown below. Stand the supports upright by
fitting the ends into holes in a large wooden or Styrofoam board.

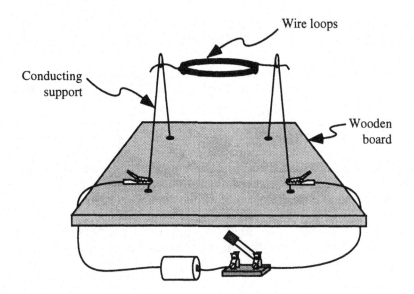

Do not close the switch yet. Close the switch only when making
observations, and then only for a few seconds at a time.

Will there be a current in the coil when the switch is closed? If so
identify the direction of the current around the loops.

While the switch is closed, which side of the coil will act like a north
pole and which side will act like a south pole? Tape a piece of paper
across the coil and label the corresponding sides "N" and "S."

C. Predict how the coil will behave when the north pole of a bar magnet is placed near the coil and the switch is closed. Explain.

Use a strong magnet to check your prediction. If the magnet is not strong enough to see any effect, then combine several magnets to make a stronger magnet. (If necessary, review the experiments in *Magnets* that involve combinations of magnets.)

D. How does the location and orientation of the magnet affect the behavior of the coil when the switch is closed? Hold the magnet near various parts of the coil and use several different orientations of the magnet to answer this question. Sketch the coil and the magnet for each case and record the behavior of the coil.

Experiment 3.2

A. Hold the north end of the magnet near one side of the coil, close the switch briefly and note the behavior of the coil.

Where can you hold the magnet, and in what orientation, so that when you briefly close and open the switch, the coil will rotate the greatest amount?

B. Develop a technique for holding the magnet in place and for opening and closing the switch that results in the coil rotating continuously. (Hold the magnet stationary and do not touch the coils.)

Account for how your method works by describing the interaction between the coil and the magnet for one complete cycle of the coil.

✔ Check your work with a staff member.

In order to make the coil in the preceding experiment rotate continuously, it was necessary for you to repeatedly open and close the switch. In the following experiment, we consider an alternative method.

Experiment 3.3

A. Obtain some enamel-coated wire.

Test to see whether the enamel is a conductor or an insulator. (If necessary, review the relevant experiments in *Electric Circuits*.)

Make a coil from the enamel-coated wire that is similar to the coil you used in Experiment 3.1.

Use sandpaper to scrape off the enamel coating from half of each of the wire leads. (See figure.) Do not scrape the loops, just the leads.

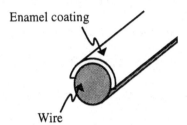

Enamel coating

Wire

B. Place the coil into the setup that you used in Experiment 3.1. Manually rotate the coil through one complete turn. Do not close the switch yet.

Imagine that the switch were closed. For which orientations of the coil will there be a current through the coil? For which orientations will there be no current through the coil? Draw sketches to illustrate your answer.

During the time that there is a current through the coil, which side of the coil will act like a north pole and which side will act like a south pole? Tape a piece of paper across the coil and label the corresponding sides "N" and "S."

C. In Experiment 3.2 you opened and closed the switch to make the coil rotate continuously.

How does one complete rotation of the coil with the leads half-stripped compare to turning on and off the switch?

Predict where you could hold a magnet so that the coil in this experiment will turn continuously.

Check your prediction. If nothing happens when you initially close the switch, give the coil a small push to get it rotating.

D. After you have the loop spinning continuously, try to make the coil spin at a faster or slower rate by changing the location or orientation of the magnet.

Sketch the position and orientation of the magnet relative to the loop to show how to hold the magnet so that the loop spins (1) at a high speed, (2) at a medium speed, and (3) at a slow speed. On each sketch, indicate the location of the north and south poles of the magnet and of the coil.

E. Investigate other changes that you think might affect the rate at which the coil spins. Some effects you might investigate include: adding a second battery in series to the first, increasing or decreasing the number of loops in the coil, and holding a second magnet near the coil.

✔ Check your work with a staff member.

The device you built in Experiment 3.3 is a form of electric motor. By using stronger magnets or larger currents, it is possible to make motors that can drive machines and perform various tasks. In some motors, the coils are held stationary and the magnet spins. If some electric motors are available in the classroom, examine them and see if you can recognize the parts that correspond to the parts of your motor.

Supplementary problems for *Electromagnets*

Note: In the supplementary problems, unless otherwise indicated, assume that the magnetic field of the earth can be neglected.

Problem 1.1

Two long, straight wires, each connected to its own battery, pass vertically through a sheet of cardboard. A compass is located the same distance from both wires at the location shown. When both switches are closed, the compass is observed to point in the direction of the arrow shown. The top of the first wire is connected to the positive terminal of a battery.

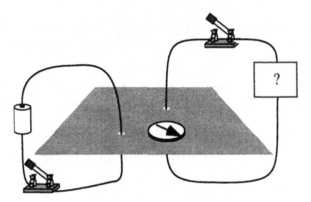

A. Is the top of the second wire connected to the positive terminal or to the negative terminal of a battery? Explain how you can tell.

B. Suppose that the orientation of the battery connected to the second wire were reversed. In which direction would the compass now point? Explain your reasoning.

Problem 1.2

A compass is placed near a current-carrying wire that is formed into the shape shown below. Determine the direction that the compass will point when placed at the three locations in the diagram. Explain. (*Hint:* Use your knowledge of the direction of the field around a single wire to help you determine the direction for a pair of wires.)

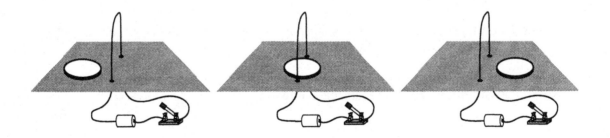

Problem 2.1

Describe how to make a strong magnet out of an initially unmagnetized iron rod without using the earth or another bar magnet. Include in your instructions the materials one would need, how one could make the magnet stronger or weaker, and where the poles of the magnet would be located.

Problem 2.2

A student has three metal bars and a coil of wire wound around a nail and attached to a battery. Each of the three metal bars might belong to one of the following categories: (i) permanent magnets, (ii) ferromagnetic objects that are not permanently magnetized, or (iii) objects made of nonmagnetic materials.

The bars and coil are numbered 1 through 4. The ends of each of these four objects are labeled *A* and *B*. The middle of each object is labeled *m*. (See figure below.)

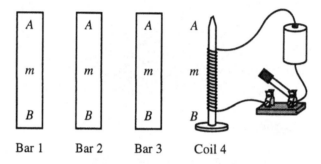

| Bar 1 | Bar 2 | Bar 3 | Coil 4 |

Assume that if a bar is a magnet, the poles are on the ends.

The student closes the switch to the coil and makes the following observations:

- 1*A* attracts 4*A*.
- 2*A* does not interact with 4*m*.
- 3*A* repels 4*B*.

A. Is bar 1 a permanent magnet? If it is not possible to tell, to which category could it belong? Explain your reasoning.

B. Is bar 2 a permanent magnet? If it is not possible to tell, to which category could it belong? Explain your reasoning.

C. Is bar 3 a permanent magnet? If it is not possible to tell, to which category could it belong? Explain your reasoning.

D. How would 1*A* interact if it were brought near 4*B*? If it is not possible to tell, what are the possibilities? Explain your reasoning.

Problem 3.1

A. The motor below is spinning in the direction indicated with the magnet in the orientation shown. Identify the north and south side of the coil of wire. Explain your reasoning.

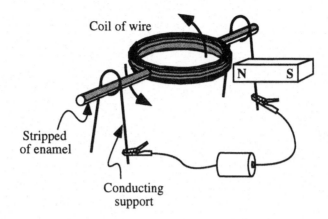

B. Suppose the magnet were oriented as shown below. At the instant shown, would the rate at which the coil is spinning increase, decrease, or remain the same? Explain.

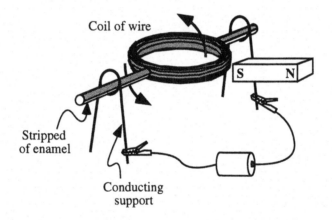

PHYSICS BY INQUIRY

Light and Optics

In *Light and Optics,* we draw on observations in the laboratory to develop a

conceptual model for geometrical optics. We use this model to predict and

explain a variety of simple phenomena and extend it to account for image

formation by mirrors, lenses, and other optical instruments.

Note: *Light and Color* (Volume I) and *Light and Optics* (Volume II) are independent. Both begin with the same concepts but subsequent topics are different. The first module places a greater emphasis on the development of basic concepts. It is recommended that students work through *Light and Color* before *Light and Optics.*

Part A: Plane mirrors and images

In Part A of *Light and Optics,* we study how light reflects from mirrors. From our observations, we construct a model that enables us to account for the formation of images in plane mirrors.

Section 1. Introduction to reflection

Experiment 1.1

This experiment should be performed in a darkened room.

Obtain a light box. (A light box is a box that has a bulb inside it and a small hole over which a mask can be placed.) Make two masks by cutting vertical slits approximately 50 mm by 3 mm in black construction paper. One mask should have one slit; the other should have two slits approximately 10 mm apart.

A. Place the light box on a large sheet of paper. Put the two masks on the light box, one at a time, and describe your observations.

What do you think would be the effect of changing the width of a slit in a mask? Make a new mask with a different slit width and check your answer.

B. Obtain two mirrors. Use blocks of wood or other objects to hold the mirrors upright as shown.

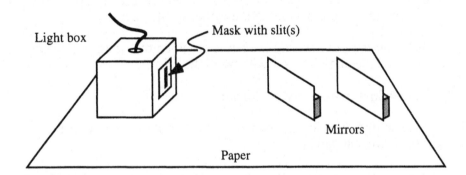

(1) Use the mask with a single slit to produce a beam of light. Examine the effect of putting one or both mirrors in the path of a beam of light. Describe your observations.

Describe the path of the light when the beam is aimed (a) at an angle toward the mirror and (b) straight toward the mirror.

McDermott & P.E.G., U.Wash./ *Physics by Inquiry*

(2)   Place the two-slit mask on the light box.

Is it possible to make the two beams cross using only one mirror? If so, draw a sketch that illustrates your answer.

(3)   Place the single-slit mask on the light box and turn off the light. Make an X on the paper in a location that is not directly in front of the slit.

Decide where you need to place one or both mirrors so that the beam will pass over the X. Talk to a staff member if you need additional equipment. Explain how you determined your answer.

Turn on the light box and check your answer.

(4)   Ask your partner to look away while you place one or two mirrors in the path of a beam of light. Mark the resulting path on the sheet of paper, then remove the mirrors and turn off the light box.

Ask your partner to predict where to place the mirrors in order to have the light follow your marked path.

Turn on the light box and check your partner's answer. Switch roles with your partner and repeat part (4).

Experiment 1.2

Place single-slit mask on the light box and put a mirror in the beam. Draw a line along the front of the mirror to mark its location.

A.   Aim the beam so it reaches the mirror at an angle.

Use a ruler to mark the path of the light both toward and away from the mirror by drawing a line along the center of the beam.

Use a protractor to measure the angle at which the beam strikes the mirror and the angle at which it leaves the mirror.

How do the two angles compare?

B.   Move the light box so that the beam strikes the mirror at a different angle and repeat part A.

C.   Summarize your observations in this experiment as a rule describing how to predict the path of a beam of light that is aimed at a mirror.

Is your rule true for all angles or only for certain angles?

Does your rule describe the behavior of each edge of the beam as well as the center of the beam?

✔ Check your rule with a staff member.

We call the beam that strikes the mirror the *incident* beam. The beam that leaves the

mirror is called the *reflected* beam.

Experiment 1.3

This experiment should be performed in a darkened room.

You will need colored construction paper, white paper, a mirror, a flashlight, and a mask for the flashlight. Make the mask by cutting a small slit (approximately 25 mm by 5 mm) in an index card or other thick piece of paper. Tape the mask to the front of the flashlight.

A. Shine the masked flashlight onto various flat surfaces, including colored pieces of construction paper.

Describe what you observe as you vary the angle between the flashlight and the surfaces.

Compare what happens when you shine the flashlight on various types of surfaces, such as dark-colored surfaces, light-colored surfaces, rough surfaces, and smooth surfaces.

B. Place a white sheet of construction paper flat on the table and hold a white sheet of paper near it as shown. Aim the flashlight at the construction paper, not the paper you are holding.

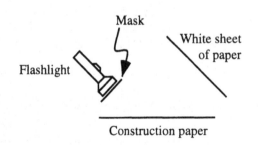

How does the appearance of the paper that you are holding change as the light is turned on and off?

How might you account for your observations?

Replace the white construction paper with red construction paper. Does this change affect the appearance of the sheet that you are holding?

C. Repeat part B using several other colors of construction paper.

How does the color of the construction paper affect what you see on the white paper you are holding? How can you account for your observations?

D. Imagine that the construction paper in part B were replaced by a mirror.

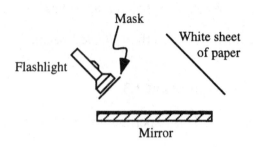

Predict whether this change would affect what you see on the paper that you are holding. Explain.

Discuss your prediction with your partner, then check your answer.

Suppose that you removed the white sheet of paper and placed your eye where the paper had been located.

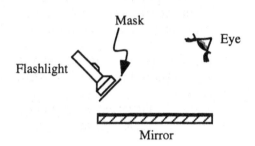

Predict what you would see if you were to look toward the mirror in this case. Explain.

Discuss your prediction with your partner, then check your answer.

E. Imagine that you were to place the sheet of white paper at the location shown at right.

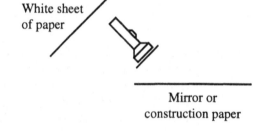

What would you see on the white sheet of paper if the flashlight were aimed at:

(1) a mirror?

(2) construction paper?

Discuss your predictions with your partner, then check your predictions.

F. Compare and contrast how light is reflected by mirrored surfaces and by non-mirrored surfaces.

Recall the rule that you wrote in Experiment 1.2 to predict the path of a beam of light that is aimed toward a mirror.

Does this rule also describe the behavior of a light beam that is reflected by a non-mirrored surface? Explain your reasoning.

✔ Discuss your results with a staff member.

McDermott & P.E.G., U.Wash./ *Physics by Inquiry*

Exercise 1.4

A. Consider the following statement made by a student:

"At night it is dark, but during the day it is bright because the sun lights up objects. When objects are lit up I can see them."

Do you agree with the statement by this student? Would you change the statement in any way to make it more complete? Base your answer on your findings from the preceding experiments.

B. Summarize the conditions necessary for you to be able to see an object.

✔ Check your reasoning with a staff member.

Experiment 1.5

For this experiment you will need a mirror, a small bulb, an index card (or other piece of heavy paper), a nail, and a large sheet of white paper.

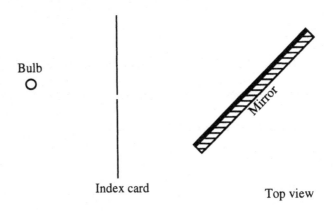

Cut a narrow slit (2–3 mm wide) in the index card. Then, on the large sheet of paper, set up the lighted bulb, slit, and mirror as shown.

A. Describe the path of the light through the slit. Sketch the path on the paper. Show the beams that are incident on and reflected from the mirror. Draw arrow heads to indicate the direction the light travels.

Where must you place your eye to see the light bulb in the mirror?

The bulb that you see in the mirror is called the *image* of the bulb.

B. Replace the bulb by a nail at the same location.

Where must you place your eye to see the image of the nail?

Describe the path that light takes from the nail to your eye.

How does the line that you drew in part A compare to the path of the light from the nail to your eye in part B?

McDermott & P.E.G., U.Wash./*Physics by Inquiry*

C. Move the index card to a
new location as shown.

Predict where you must
place your eye to see the
image of the nail. Explain.   Nail

Check your prediction.
On the paper, mark the
path that light takes from
the nail to your eye. Draw
arrow heads to indicate the
direction the light travels
along the path.

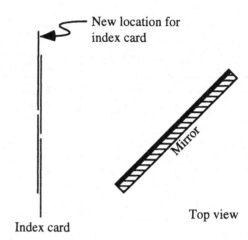

D. Remove the index card. Find the region in which you must place your eye in
order to see the image of the nail. Draw lines on the paper to show the path of
the light from the bulb to your eye at the extreme limits of the range.

In the preceding experiment, you drew lines with arrow heads (⟶———) to
indicate the path of light. These lines are called *rays*. A diagram in which rays are used
is called a *ray diagram*.

Exercise 1.6

The diagram below shows an object near a mirror. An observer is free to
walk along the line in front of the mirror.

A. From which of the lettered points along the line could the observer see
the image of the object? Explain.

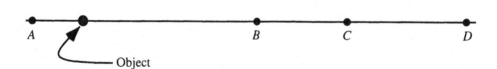

B. For each point from which the observer can see the image, indicate the
direction in which the observer must look in order to see the image.

Experiment 1.7

Ask a staff member to show you the demonstration for this experiment.

A. Write a description of the demonstration in your notebook. Record your predictions.

B. Observe the demonstration. Use a ray diagram to account for your observations.

Exercise 1.8

Consider a pencil near a mirror as shown in the top view diagram below.

Mirror

Pencil

A. Indicate the region in front of the mirror where an observer could see:

   (1) the image of the tip of the pencil

   (2) the image of the eraser

   (3) the entire image of the pencil

B. Choose a point from which the observer could see the entire image of the pencil. Indicate the directions in which that observer would have to look to see:

   (1) the image of the tip

   (2) the image of the eraser

C. Place a mirror on your ray diagram and check your results for parts A and B. (If necessary, use paper to mask a larger mirror to make it the correct size.)

✔ Discuss this experiment with a staff member.

McDermott & P.E.G., U.Wash./ *Physics by Inquiry*

Section 2. Image formation in a plane mirror

In this section, we investigate image formation in a plane mirror. We develop techniques that can be used to determine the location and size of an image.

Experiment 2.1

A. Close one eye and place your open eye at table level. Have your partner drop a small piece of paper (1 cm by 1 cm) onto the table.

Hold your finger above the table and then move your finger until you think it is directly above the piece of paper. Move your finger straight down and see if it was actually directly above the paper.

Try this procedure several times, with your partner dropping the piece of paper at different locations. Keep your open eye at table level. After several tries, switch roles with your partner.

Were you and your partner consistently able to locate the piece of paper?

How can your account for the fact that when your finger misses the piece of paper, your finger is in front of the paper or behind it, but not to the left or right of the paper?

B. Try to develop a strategy that allows you to consistently locate the piece of paper with your finger.

Experiment 2.2

A. Have your partner hold two pencils one above the other about one meter in front of you. Then have your partner move one pencil about 15 cm closer to you.

Close one eye, then move your head so that one pencil appears to be directly above the other. Now move your head from side to side. Describe what you observe.

B. Again, close one eye and move your head so that the pencils appear one directly above the other.

If you move your head to the right, which pencil appears to be on the right: the one closer to you or farther from you?

If instead you move your head to the left, which pencil appears to be on the left: the one closer to you or farther from you?

C. Have your partner move the pencils closer together.

    How does this change affect what you observe when you move your head from side to side?

    Have your partner place the two pencils one above the other.

    How does this change affect what you observe when you move your head from side to side?

In the previous experiment, you observed that there is an apparent change in the relative location of the two pencils when you move your head from side to side. This effect is said to be due to *parallax*.

Experiment 2.3

A. Describe how you can use parallax to determine which of two objects is closer to you. Draw a ray diagram to illustrate your answer.

    How must the two objects be located relative to one another if you observe no effect of parallax when you move your head from side to side?

B. Explain how you could use parallax in Experiment 2.1 to tell whether your finger is (1) directly over the piece of paper, (2) in front of the piece of paper, or (3) behind the piece of paper.

    Repeat Experiment 2.1 and test the method that you have devised.

C. Have your partner hold two pencils in front of you as in part A of Experiment 2.2. Close one eye and use the method of parallax to direct your partner as to which direction to move the upper pencil until it is directly above the lower pencil.

✔ Discuss your reasoning with a staff member.

Experiment 2.4

Obtain two identical nails and a mirror that is shorter than the nails. Place one nail upright on a sheet of paper about 10 cm in front of the mirror. We will call this nail the "object nail."

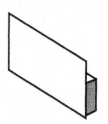

A. Place your head so that you can see the image of the object nail. Then move your head from side to side. Describe what you observe.

> Where does the image appear to be located? Does the image stay in the same location or does it change location as you move your head? If there is a single location, mark that location on the paper. If the image location depends on where you place your eye, mark several locations of the image for several different eye locations.

B. Place your head so that you can see the image of the object nail. Apply the strategy for using parallax that you developed in Experiment 2.3 to place the second nail at the location of the image of the object nail. Mark this location on the paper.

Move your head to a new location and repeat the procedure in part B.

> Is there a unique location where all observers would agree that the image is located (provided they can see an image)? Check your answer experimentally.

C. Two students discuss the image of a nail in a mirror:

Student 1: *"The location of the nail's image changes as I move my head. When I move to the left, the image moves to the left side of the mirror. When I move to the right, it moves to the right side of the mirror. The image location is changing."*

Student 2: *"Your observations tell you that the image isn't located on the surface of the mirror. The image acts like it is behind the mirror. Since the image seems to follow you, as compared to the mirror, the image must be behind the mirror."*

Do you agree with student 1, student 2, or neither? Explain your reasoning.

✔ Discuss this experiment with a staff member.

In Experiment 2.3, you developed a strategy for finding the location of an object using the effect of parallax. In Experiment 2.4, you applied your method to find the location of an image. We refer to the process of using parallax to determine the location of an image as the *method of parallax* or *locating an image by parallax.*

In the following experiments, we determine the location of an image by another technique called *ray tracing.* This technique is based on our model for light in which we envision light as being emitted in all directions by luminous objects, such as light bulbs, or as being reflected in all directions by non-luminous objects, such as nails or pins.

Experiment 2.5

Secure a sheet of paper to a piece of corrugated cardboard. Stand a nail on its head at one end of the paper.

A. Place your head at table level at the other end of the paper from the nail. Close one eye and look at the nail.

Push two pins into the cardboard so that, from your location, they appear to be in line with the nail.

The two pins determine your line of sight to the nail. Use a ruler to draw your line of sight to the nail.

Do you determine your line of sight more accurately by placing the pins close together or far apart?

Repeat the procedure above to mark lines of sight from three other vantage points, then remove the nail.

How can you use the lines of sight that you drew to determine where the nail was located?

What is the smallest number of lines of sight necessary to determine the location of an object?

B. Turn the paper over (or obtain a fresh sheet). Place a mirror and nail on top of the paper as shown at right. Draw a line on the paper to mark the location of the mirror.

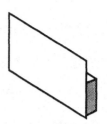

(1) Place your head near the surface of the table and look at the image of the nail. Push two pins into the cardboard so that, from your location, they appear to be in line with the image of the nail. Use a ruler to draw your line of sight to the image.

Can you tell the location of the image from the single line of sight that you have drawn? Explain.

Repeat the procedure above from several different eye locations.

How can you use the lines of sight that you drew to determine the location of the image of the nail?

Would *all* observers who can see the image of the nail agree on its location? Explain.

(2) Use the method of parallax to determine the location of the image of the nail.

Does the method of parallax yield the same image location as you found in part (1) above?

(3) Discuss how the methods in part (1) and part (2) for determining image location are related.

C. Move the nail and use the method developed above to find the new location of the image. Repeat for several different nail locations including: a location close to the mirror, a location far from the mirror, and a location off to the side of the mirror. Use a different sheet of paper for each case.

D. Based on your observations, describe the relationship between the object location, the image location, and the location of the mirror.

You will need the sheets of paper from part C for the next experiment.

Experiment 2.6

A. For this part of the experiment, use one of the sheets of paper from part C of Experiment 2.5. For each of your eye locations in that experiment, draw a ray that shows the entire path of light from the object nail to your eye. Base your rays on the lines of sight that you drew to the image of the nail. Draw an arrow head on each line segment to show the direction of the light.

B. Imagine that several observers are looking through a hole in the wall at a nail.

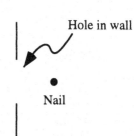

Draw several rays from the nail to various observers. Use arrow heads to indicate the direction of the light.

C. Compare your ray diagrams for parts A and B above. Discuss how the two situations are similar. In particular, to which feature in Experiment 2.5 would the nail correspond? To what would the hole in the wall correspond?

D. An object is placed near a mirror as shown.

(1) On the diagram, draw a *solid line* to represent one ray of light from the object to the mirror. Show the path of the reflected light also. (Use a protractor to find the reflected ray.)

> If you looked back along the reflected ray, what would you see?

> What can you tell about the location of the image from the single ray you have drawn?

> Does a single ray *alone* give you enough information to determine the location of an image?

(2) Repeat part (1) for a second ray from the object to the mirror.

> Describe how to use the two rays you have drawn to determine the image location. Show the image location on your diagram.

Note: When extending a ray behind a mirror, it is conventional to use a *dashed line,* rather than a solid line. This helps distinguish between a path that light actually took and a path that light *appears* to have taken.

> What is the smallest number of rays you must draw when using a ray diagram to locate the image of an object?

✔ Discuss your results with a staff member.

     McDermott & P.E.G., U.Wash./*Physics by Inquiry*

In this course, when you are asked to use ray tracing to determine the location of an image, always draw at least two rays. Use a solid line *with an arrow head* (——➤——) to represent the actual path of a ray of light. Use a dashed line (— — — — — — —) to extend a ray behind a mirror to indicate a path that light only *appears* to take.

Exercise 2.7

As you may have noticed in your experiments thus far, the image of an object in a plane mirror is located as far behind the mirror as the object is in front. Use geometry to prove this fact. (*Hint:* Start with a ray diagram that shows the location of an image and look for similar triangles. You may find it necessary to draw additional lines or rays.)

Exercise 2.8

A. Draw a ray diagram to determine the location of the image of each of the two objects in the diagram below.

B. Determine the region in which you could stand and see the images of both objects at the same time.

C. Describe how you could use a ray diagram to determine the location of the image of an extended object, such as a pencil or brick.

Exercise 2.9

Draw a ray diagram to find the location of the image of the nail shown in the side view diagram below.

     Nail   Mirror

Experiment 2.10

Obtain a half-silvered mirror and two identical cylinders (e.g., cans or batteries). Examine the half-silvered mirror and describe how it differs from a standard mirror.

A. Place one cylinder (the "object cylinder") in front of the mirror. While looking at the image of the cylinder, move the cylinder away from the mirror.

How is the size of the image affected by moving the cylinder away from the mirror?

B. Return the object cylinder to its original position in front of the mirror. Use the method of parallax to place the second cylinder at the location of the image of the object cylinder.

How does the size of the image compare to the size of the object cylinder? Explain your reasoning.

Move the object cylinder farther away from the mirror and repeat the experiment.

Is the size of the image still the same as it was before? Explain how you can tell.

C. Try to resolve any conflict in your answers to parts A and B.

✔ Check your reasoning with a staff member.

         McDermott & P.E.G., U.Wash./*Physics by Inquiry*

Experiment 2.11

A. Imagine that you are looking at yourself in a mirror mounted on a wall. From where you stand, you are able to see from the top of your head down to the top of your shoulders.

Suppose you were to back away from the mirror. Would the amount of your body that you see in the mirror increase, decrease, or stay the same? Explain. Include a diagram as part of your explanation.

B. What is the minimum size mirror you need to view your entire body? Draw a ray diagram that supports your answer.

(1) Does moving closer to or farther from the mirror change how much of yourself is visible in the mirror?

(2) Does moving closer to or farther from the mirror change the size of your image?

(3) Imagine that your eyes were at the top of your head. Would your answers to parts (1) and (2) above be different in this case?

Draw ray diagrams to justify your answers to the questions above.

C. If possible, perform an experiment to test your answers to the questions above.

D. Does moving closer to or farther from a mirror change how much of the room is visible in the mirror? Draw a ray diagram(s) that supports your answer.

✔ Check your diagrams and your reasoning with a staff member.

Exercise 2.12

Discuss how the size of an object compares to the size of its image in a plane mirror. Describe how, if at all, the size of the image would change if (1) the object were moved farther away from the mirror, or (2) you, the observer, were to move farther away from the mirror.

Section 3. Multiple images

Experiment 3.1

Obtain two mirrors and a small object, such as a coin.

How many images of the coin can you see using only one mirror?

How many images can you see using two mirrors?

Examine the ways in which you can use two mirrors to make more than one image. As you work , try to think of questions and attempt to answer them by designing and performing experiments.

As part of your explorations, try to answer the following questions:

What is the greatest number of images of the coin that you can make? How are the mirrors arranged to do this?

Can you arrange the two mirrors so that there are only two images of the coin? If so, how should the mirrors be arranged?

Does the number of images that you see depend on where you place your eye? Is this true for all arrangements of the mirrors?

Are all of the images identical to one another and to the coin? Describe any differences you see.

Can you make two images overlap so that they appear to be a single image? Can you do this when the two overlapping images are identical? Can you do this when the two overlapping images are different? How many images would you say that there are in each case?

Can you make an even number of images? Can you make an odd number of images?

How many images can you see if the mirrors are placed parallel to each other but on opposite sides of the coin?

Experiment 3.2

A. The top view diagram at right shows a line drawn on a piece of paper and a mirror placed on top of the line. Draw a sketch that shows what you think you would see in the mirror.

Obtain a mirror and check your prediction.

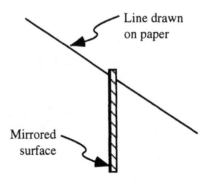
Line drawn on paper

Mirrored surface

McDermott & P.E.G., U.Wash./*Physics by Inquiry*

B. Place a narrow piece of tape along the bottom edge of each of two mirrors as shown below. Make each of the tapes a different color in order to be able to tell the mirrors apart.

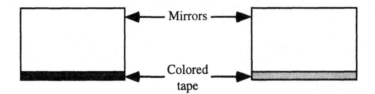

Stand the mirrors upright on a piece of paper with the taped edges on the bottom. Place the mirrors side-by-side along the 0°–180° line on a sheet of polar graph paper. Slowly rotate the mirrors so that the angle between them is a little less than 180°. Describe what you observe.

How does what you see in the mirrors relate to what you observed in part A?

C. Slowly decrease the angle between the mirrors, keeping the junction of the mirrors centered on the paper.

What happens when the angle between the mirrors reaches 90°?

What happens when the angle becomes a little less than 90°?

Continue to decrease the angle. Describe your observations.

D. Set up the mirrors so that you can see at least two images of each mirror. On another piece of polar graph paper, draw a diagram that shows the arrangement of the mirrors and the images.

On the diagram, indicate which angles are equal.

Is it possible to arrange the mirrors so that all the adjacent angles in your diagram are equal? Describe the arrangement(s) of the mirrors for which this would happen.

E. In the situation that you sketched in part D, are there any images that appear to be the image of another image?

If so, how can you use the idea of an "image of an image" to account for all of the images of each mirror?

Experiment 3.3

A. Each of the diagrams below shows two mirrors held at an angle. Using a straightedge and protractor, copy each of the diagrams onto a separate sheet of paper.

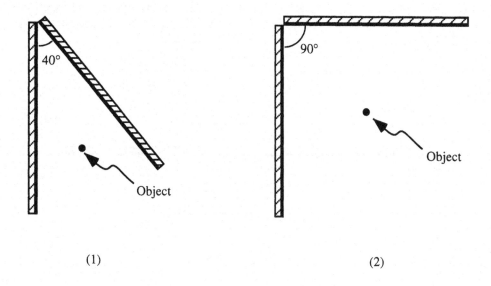

(1)   (2)

The idea of an "image of an image" can be used to predict the number of images and the image locations. Apply this idea in each case above to:

- Determine the number of images of the mirrors and their locations.

- Determine the number of images of the object and their locations.

Imagine that one of the mirrors in each figure were removed. Predict which image(s) you would still see and which image(s) would vanish.

Check your predictions. If a prediction was incorrect, try to resolve the conflict before continuing.

B. In your own words, describe the approach that you used in part A to determine the number of images and the image locations. Your description should be sufficiently clear that someone else could read it and apply it to a new situation.

✔ Check your answers for this experiment with a staff member.

         McDermott & P.E.G., U.Wash./ *Physics by Inquiry*

Experiment 3.4

Remove the tape from the mirrors you used in the preceding experiment. Place the mirrors in contact face-to-face. Tape one edge of the mirrors together so that when they are opened, the mirrors are held together.

On a separate sheet of paper, draw an *X* and a dot as shown in the diagram below. Place the mirrors on the paper with the hinge between the mirrors on the *X* as shown.

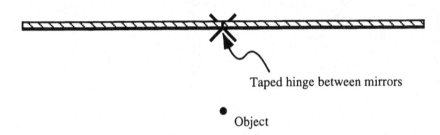

Taped hinge between mirrors

● Object

In this experiment, we *qualitatively* examine how the angle between the mirrors affects what you see in the mirrors.

A. Gradually close the mirrors, all the while keeping the hinge on the *X* and keeping the object midway between the mirrors.

Describe what you observe as the mirrors are closed. Notice how the number of images of the object changes.

B. On the paper beneath the mirrors, mark the locations of the mirrors when there is one image, two images, three images, and so on.

In which case is the number of images more sensitive to a small change in the angle between the mirrors: when the angle is large or when the angle is small?

C. For certain angles between the mirrors, both the object and one of the images lie on the bisector of the angle formed by the mirrors. In this case we call the image a *central image*.

For which angles is there a central image?

What is distinctive about the number and arrangement of images when there is a central image?

✔ Check your results with a staff member.

McDermott & P.E.G., U.Wash./ *Physics by Inquiry*

Experiment 3.5

Copy each of the diagrams below onto a separate sheet of paper. Use a straight-edge and a protractor.

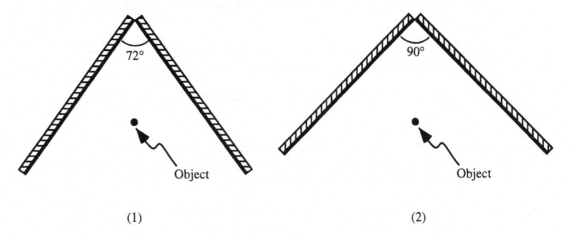

(1)   (2)

Arrange two mirrors and an object (e.g., a nail) as shown in each of the diagrams. For each case, answer the following questions.

A. Use the method of parallax to find the locations of the all the images of the object. Mark these image locations on your diagram.

   How would you characterize the arrangement of the images?

B. Find the locations of the images of the mirrors. Mark these locations on your diagram.

   Indicate which images of the mirrors can be seen in each mirror. (Are there any that can be seen in both mirrors?)

   How would you characterize the locations of the images of the mirrors in each figure?

C. For each figure, how are the locations of the images of the mirrors related to the locations of the images of the object?

Experiment 3.6

In this experiment, we determine a mathematical relationship between the number of image locations and the angle between the mirrors.

Place two mirrors along the 0°–180° line of a sheet of polar graph paper. The hinge between the mirrors should be at the center of the paper. Put a pin on the 90° line. Slowly close the mirrors, keeping the pin in the center between the mirrors.

McDermott & P.E.G., U.Wash./ *Physics by Inquiry*

A. Make the measurements necessary to complete the table below. Note that in some cases, there is a range of angles for which a given number of images are visible. For the purpose of this experiment, you should pick a single angle to represent each range. The following questions can help you decide on a criterion for picking a particular angle.

(1) For which of the cases listed in the table does a single mirror angle correspond to a particular number of images?

How is the arrangement of images similar for these cases?

(2) For each case in which a range of angles corresponds to a particular number of images, is there a single angle that results in a similar arrangement of images as in (1) above?

What criterion might you use to pick one specific angle to characterize a range of angles? What reason can you give for choosing that particular angle rather than some other angle?

Mirror angle	Number of image locations (given as a single angle)
	1
	2
	3
	4
	5
	6
	7
	8
	9
	10

B. Use the data above and your diagrams in Experiments 3.3 and 3.5 to infer a mathematical relationship between the number of image locations and the corresponding angle between the mirrors.

✔ Check your reasoning with a staff member.

Experiment 3.7

A. An object is placed on the bisector of the angle between two mirrors. Three images result. The images and the object are equally spaced.

What is the angle between the mirrors? Explain.

Draw a diagram showing two mirrors with the appropriate angle between them. Use ray tracing to determine the image locations.

Is there a central image in this case? If so, how many times is a light ray reflected in producing the central image?

How many times is a light ray reflected in producing each of the other images?

B. At what angle should the mirrors be placed so that there are two images of the object, and the images and object are equally spaced?

Use ray tracing to determine the location of the images.

Use your ray diagram to explain why there is not a third image.

Experiment 3.8

Find a coin that has a head engraved on one side of it.

Does the coin show the left or the right side of the person's head?

Stand the coin upright in front of a single mirror so that the image of the head is upright. Which side of the person's head do you see on the image?

The image of the head is said to be a "mirror image" or a *perverted* image.

Experiment 3.9

A. Place a hinged set of mirrors on the 0°–180° line of a piece of polar graph paper. Place a coin on the 90° line in front of the mirrors. As you close the mirrors note whether the emerging central image is perverted. Record the pattern of perverted images and non-perverted images for several cases, including some with a central image and some for equally spaced images without a central image. Describe this pattern in words.

B. Describe the relationship between the number of times a light ray is reflected in producing an image and whether or not that image is perverted.

✔ Check your answers with a staff member.

Part B: Lenses, curved mirrors, and images

In Part B of *Light and Optics,* we study the behavior of light as it passes from one material to another, such as from air to glass or from air to water. We develop a technique for drawing ray diagrams that enables us to determine the location and size of an image formed by a lens. We extend the model to account for the images formed by curved mirrors and apply it in constructing optical instruments, such as a simple telescope and microscope.

Section 4. Introduction to refraction

Experiment 4.1

This experiment should be performed in a darkened room.

A. Obtain a light box and two clear glass beakers. Fill the beakers with water to a level that is higher than the slits of the light box.

Place the light box and beakers on a large sheet of white paper. Turn on the light box and explore what happens when one or both of the beakers are placed into a beam of light.

Describe some of your observations, both in diagrams and in words. Ask a staff member for assistance if you need additional materials for further explorations.

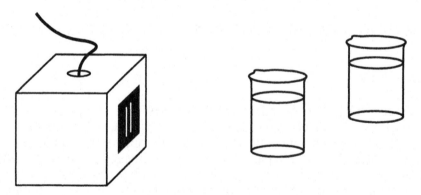

On the white paper and in your notebook, record several arrangements of the beakers and light beams. Try both single and double beams.

If you can see the path of a beam through the water, record that also. If you cannot see the path, can you infer it based on where the light entered and exited the beaker?

B. As part of your explorations, try to answer the following questions:

Does light follow a straight path in water the way it does in air?

How does the path of a beam that passes through a beaker of water differ from the path through an empty beaker?

Can you use a single beaker of water to make the beam bend:

(1) left then left again?

(2) left then right?

(3) right then right again?

(4) only once either left or right?

(5) not at all?

Describe how to place a beaker of water to produce (1) the largest overall bend in a beam and (2) the smallest overall bend in a beam.

Does changing the liquid in the beaker affect the behavior of the light? If so, how? Try various liquids. In each case decide whether the effect is greater than or less than it is with water.

C. On the basis of your observations, would you say that it is the glass from which the beaker is made or the liquid in the beaker that is responsible for the bend in the beam? Explain.

D. Describe in words what happens to a beam when it strikes a beaker of water at various angles.

Experiment 4.2

Repeat Experiment 4.1 using containers with straight sides rather than curved sides. Record your observations.

Compare and contrast your results with the results of Experiment 4.1. How is the behavior of light similar for the two types of containers, and how is it different?

✔ Check your results from Experiments 4.1 and 4.2 with a staff member.

We call the bending of light when it passes from one material to another (e.g., from air to water) *refraction*. This word comes from the Latin *refringere, refractus:* to break off. We say that light *refracts* in passing from one material to another. The beam in the second material is called the *transmitted* or *refracted* beam. As you may have observed, often light is both transmitted and reflected at the boundary between two materials.

Exercise 4.3

The beam inside a beaker may be considered as both a transmitted beam and an incident beam: it is the transmitted beam where light enters the beaker and it is the incident beam where light leaves the beaker.

Examine several of your diagrams from Experiments 4.1 and 4.2. For each diagram, identify the incident, transmitted, and reflected beams. Clearly label the beams that are both incident and transmitted beams.

Experiment 4.4

A. Obtain two flat sheets of glass, one thick and one thin (e.g., a microscope slide and a piece of window glass).

Place each piece of glass, in turn, into a beam of light and examine the effect on the beam. Try various orientations of the glass. Sketch diagrams to record your observations.

Does one of the pieces of glass seem to affect the beam to a greater extent? If so, which one? Describe the experiment(s) that allow you to determine your answer.

B. In Experiment 4.1 you examined the path of light through a beaker of water. You determined whether the glass from which a beaker is made plays a large role in refracting light.

Are your observations in part A above consistent with your conclusion in Experiment 4.1? Explain.

Exercise 4.5

In each top view diagram below, the path of a beam of light is shown incident on the side of a transparent container of water. Only part of each container is shown.

Sketch the approximate path of the beam in the second material. Using a dashed line, also draw the path that the beam would have taken had it continued without bending. Be sure that your rays are consistent with your observations. Make additional observations if necessary.

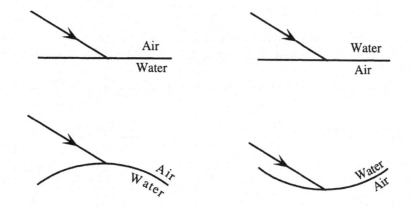

Exercise 4.6

A. Each of the top view diagrams below illustrates two rays of light incident on a container filled with water.

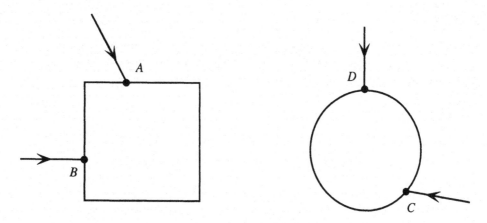

Make a sketch that shows the approximate directions of the corresponding transmitted rays. Draw the rays so that they are consistent with your prior observations.

B. Draw a line through point *A* that makes a 90° angle with the side of the container. Repeat for point *B*.

These lines are said to be *perpendicular* or *normal* to the container. Each line is called the *normal* to the container at the point it intersects the container.

Draw the normal at point *C* and at point *D* on the beaker. Explain what it means for a line to be normal to a curve.

Where do the lines that you drew normal to the circle at points *A*, *B*, *C*, and *D* intersect? Explain. Describe a simple method for drawing the normal to a circle at a given point.

C. Compare the path of each transmitted beam that you drew in part A with the path the beam would have taken if the container had been empty.

Does light bend "toward" or "away from" the normal when passing from air to water?

Does light bend in the same way for a flat or curved surface?

D. If you have not already done so, extend the transmitted rays in part A so that they reach another point on the container and pass from water back into air.

Does light bend "toward" or "away from" the normal when passing from water to air?

Does light bend in the same way for a flat or curved surface?

✔ Check your results with a staff member.

Experiment 4.7

In Section 1, you were instructed to draw directional arrows on the rays in your diagrams. Is it possible to tell the direction of a light ray from the path alone? Consider the following cases:

(1) a light ray in air

(2) a light ray reflected from a mirror

(3) a light ray refracted through two different materials

A. Design and perform experiments that allow you to check your responses. Are the results consistent with your answers above? Explain.

B. What generalization might you draw about the reversibility of a light ray? Explain.

✔ Discuss your results with a staff member.

Exercise 4.8

The top view diagrams below show beams of light incident on transparent containers of water. The beams are all incident on the containers at the same angle.

Sketch the transmitted beams in each case and show the direction of each beam when it passes back out of the containers into the air.

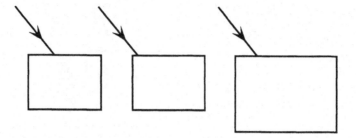

How do the directions of the beams inside the containers compare? Explain your reasoning.

How do the directions of the beams once they have left the containers compare? Explain your reasoning.

In each case, how does the direction of the beam that has exited the container compare to the direction of the beam incident on the container?

Obtain several containers of water and check your answers.

Experiment 4.9

A. Obtain a prism and place it into a beam from your light box. Try various orientations of the prism. Describe your observations.

   Does the prism refract light? In your notebook, make a sketch of the path of the light through the prism.

B. Use a two-slit mask on your light box. Place a beaker of water into the beams and adjust the beaker so that the two beams cross.

   Place a blue acetate over both slits and mark the point where the beams cross. Replace the blue acetate by a red acetate and again mark the intersection point of the beams.

   Describe your observations.

   What can you conclude about the amount by which red and blue light bend when passing from air to water? If they bend by different amounts, which bends more?

C. Discuss how your observations in part B can help you to account for your observations in part A. (You may also want to refer to the experiments on color in *Light and Color* in Volume I.)

Section 5. Law of refraction: Snell's law

Experiment 5.1

This experiment should be performed in a darkened room.

Place a semi-cylindrical dish of
water onto a sheet of paper, and aim
a beam of light from a light box at
the dish. (If necessary, raise the
dish to the height of the slits.)

Use paper to block the part of the
beam above the water surface.

A. View the dish from above and slowly rotate it through 360°. Describe
   your observations.

   Make several sketches in your notebook for various configurations
   of the dish and incident beam. On your diagrams show not only
   the incident beam, but any reflected and transmitted beams.

B. For this part of the experiment, orient the dish so that the beam always
   strikes the flat side of the dish first.

   (1) Examine what happens when the beam strikes the dish at different
       locations along the flat side and at different angles.

       Sketch various arrangements of the dish and beam.

   (2) Find orientations of the beam and dish in which the beam is first
       incident on the flat side of the dish and behaves as follows:

       • the beam bends at both the flat side and the curved side of
         the dish

       • the beam does not bend at either side of the dish

       • the beam bends at the flat side of the dish, but not at the
         curved side

       • the beam bends at the curved side of the dish, but not at the
         flat side

   Record several examples of each case (unless a particular behavior
   occurs for only one orientation of the incident beam).

C. For this part of the experiment, orient the dish so that the beam always strikes the curved side of the dish first.

Examine what happens when the beam strikes the dish at different locations along the curved side and at different angles.

Sketch various arrangements of the dish and beam.

Compare your diagrams for parts B and C. Describe the similarities and differences.

D. Orient the dish so that the beam strikes the curved side of the dish first. Gradually rotate the dish, keeping the beam aimed at the same location on the curved side of the dish.

Find an orientation of the dish for which one of the incident beams (either the beam in air or the beam in water) produces no refracted beam, that is, the beam is entirely reflected. In your notebook, sketch this configuration of dish and light beams.

E. Rotate the dish so that the beam strikes the flat side of the dish first and repeat part D.

We use the term *angle of incidence* to refer to the angle between an incident beam and the normal to the surface. The term *angle of refraction* is used for the angle between the refracted beam and the normal. The *angle of reflection* is defined in a similar way. As you have already observed, for any two materials (e.g., air and water) the angle of refraction depends on the angle of incidence. Under certain conditions, there is no refracted beam and the beam is entirely reflected. This phenomenon is called *total internal reflection*.

In the remainder of this section we examine the relationship between the angle of incidence and the angle of refraction for light passing from one material to another.

Experiment 5.2

Devise and perform experiments to answer the following questions. Record your results in your notebook.

A. Can total internal reflection occur when a light beam in water is incident on air?

Can total internal reflection occur when a light beam in air is incident on water?

B. Does total internal reflection occur for a single angle of incidence or for a range of angles? Find the angle or range of angles.

✔ Check your results from Experiments 5.1 and 5.2 with a staff member.

Experiment 5.3

In this experiment we conduct a qualitative investigation of the relationship between the angle of incidence and the angle of refraction for a light beam that passes from air to water.

A. The top view diagram at right shows a semi-cylindrical dish of water. Sketch the path of a beam of light that strikes the flat side of the dish and passes through the dish without bending at either side of the dish.

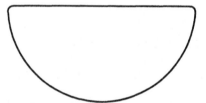

Clearly label the angle of incidence and the angle of refraction for the beam at both the flat and the curved surfaces. (*Hint:* Recall from Exercise 4.6 your method for drawing the normal to a circle at any point on the circle.)

B. Aim a beam from your light box at a semi-circular dish of water at the angle you illustrated in part A.

Gradually increase the angle of incidence, keeping the beam aimed at the same point on the flat side of the dish.

Describe your observations.

As the angle of incidence increases, does the angle of refraction increase, decrease, or stay the same?

Experiment 5.4

In this experiment we conduct a quantitative investigation of the relationship between the angle of incidence and the angle of refraction for a light beam that passes from air to water.

A. Place a semi-cylindrical dish on a sheet of polar graph paper with the center of the flat side at the center of the paper. Arrange the light beam and dish as you did in part B of the preceding experiment. Concentrate on the incident and refracted beams at the flat side of the dish.

    (1) If the angle of incidence is 0°, what is the angle of refraction?

    (2) Aim the incident beam at the center of the flat side so that the angle of incidence is 10°. Measure and record the angle of refraction for this case.

        When the angle of incidence is doubled to 20°, does the angle of refraction double? Check your answer experimentally.

    (3) Double the angle of incidence twice more to 40° and 80°. Measure and record the angle of refraction each time.

        Is the angle of refraction directly proportional to the angle of incidence for all angles of incidence? Explain your reasoning.

B. Vary the angle of incidence while keeping the incident beam aimed at the center of the flat side of the dish. Record each angle of incidence and the corresponding angle of refraction in the table at the right. Use angles of incidence ranging from 0° to 90° in increments of 10°.

Light passing from air to water

Angle of incidence	Angle of refraction
——	——
——	——
——	——
——	——
——	——
——	——
——	——
——	——
——	——
——	——

In the following experiment, we investigate an alternative and more precise method of measuring the angles of incidence and refraction.

Experiment 5.5

There is a small vertical mark at the center of the flat side of the semi-cylindrical dish. We will use that mark in this experiment.

Suppose that a pin were placed upright near the dish of water as shown.

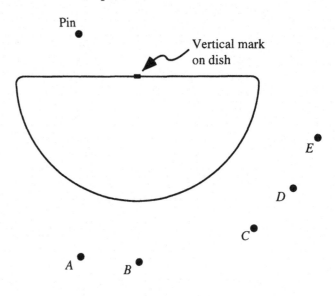

A. Imagine that you were to look at the pin through the water from each of the five locations labeled *A–E*.

> From which locations, if any, do you think the vertical mark on the dish would appear to be in line with the pin? Explain your reasoning.

Obtain a pin and check your predictions. Resolve any inconsistencies.

B. Secure a piece of polar graph paper to a piece of corrugated cardboard. Place the dish on the paper with the vertical mark on the dish at the center of the paper.

Place a pin about 4 cm from the center of the dish and at an angle of 20° from the normal to the center of the flat side of the dish.

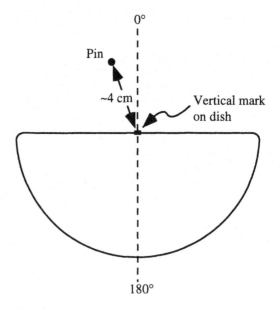

Predict where you would place your eye so that you could look through the curved side of dish and see the pin in line with the vertical mark on the dish. (*Hint:* How could you use your results from Experiment 5.4 in making this prediction?)

Check your prediction experimentally. Resolve any inconsistencies.

C. Use your results from part B to devise a simple method for determining the angle of refraction for a given angle of incidence.

Check your method by comparing the results you obtain with this method to those that you obtained in Experiment 5.4.

✔ Check your results from Experiments 5.4 and 5.5 with a staff member.

Experiment 5.6

Reverse the orientation of the semi-cylindrical dish on the polar graph paper as shown at right.

A. Devise a method for using the equipment in Experiment 5.5 to measure the angle of refraction for various angles of incidence for light passing from water to air.

(*Hint:* How could you modify your method from part C of Experiment 5.5 to do this?)

> Explain why your method gives correct values of these angles for light passing from water to air when the pin is in air, not in water.

B. Use the method you devised in part A to complete the table at right.

C. Compare the results you obtained in this experiment to those you obtained in Experiments 5.4 and 5.5. Describe any similarities and differences.

D. Earlier, you found that under certain circumstances light is totally internally reflected. Use your method from part A to find the range of angles for which total internal reflection occurs.

✔ Check your results from this experiment with a staff member.

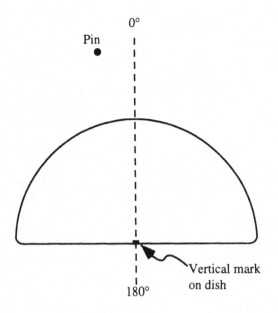

Light passing from water to air

Angle of incidence	Angle of refraction
――――	――――
――――	――――
――――	――――
――――	――――
――――	――――
――――	――――
――――	――――
――――	――――
――――	――――
――――	――――

Exercise 5.7

Suppose you were to use your data from Experiments 5.4 and 5.5 to graph angle of incidence versus angle of refraction for light passing from air to water.

Would you expect the graph to be a straight line? Explain.

Graph your data and check your prediction.

✔ Check your answers with a staff member.

Exercise 5.8

Your results so far indicate that the relationship between the angle of incidence and the angle of refraction is more complicated than just a linear one. Since we are dealing with angles, let us look for an algebraic relationship between these angles that involves trigonometric functions such as sin θ, cos θ, and tan θ.

A. Use your results from Experiments 5.4 and 5.5 to make a graph of sin θ_i versus sin θ_r, where θ_i represents the angle of incidence (in air) and θ_r represents the angle of refraction (in water). (Add additional columns to your table from Experiment 5.4 to help you make this graph.)

Should your graph go pass through the origin *(0,0)*? Explain.

Is the relationship between sin θ_i and sin θ_r linear? Explain how you can tell from your graph.

Determine the slope of the graph and write an equation that relates the slope, θ_r, and θ_i.

B. Repeat part A using the angles of incidence (in water) and angles of refraction (in air) that you listed in the table in Experiment 5.6.

How is the slope of the graph from part A related to the slope of the graph from part B?

✔ Check your answers with a staff member.

As your graphs suggest, there is a simple relationship between the sine of the angle of incidence and the sine of the angle of refraction for light that passes from one material to another. The slope of the graph of sin θ_i versus sin θ_r from part A of Exercise 5.8 is called the *index of refraction of water with respect to air*. This relationship is called the *law of refraction* or *Snell's Law*.

Experiment 5.9

A. Use the methods you have developed in this section to determine the index of refraction with respect to air of a substance other than water (e.g., glass, plastic, vegetable oil, corn syrup).

B. Describe how the index of refraction for light passing from one material to a second is related to the index of refraction for light passing from the second material to the first.

✔ Discuss your results with a staff member.

Section 6. Examples of refraction in everyday life

In this section, we investigate simple examples of refraction in everyday life. The

emphasis is on relating ray diagrams to real world observations.

Experiment 6.1

Place a coin at the bottom of an empty can (or a beaker that has the sides covered so that you cannot see through them).

Look into the can at the coin while your partner slowly moves the can away from you until you no longer see the coin. Keep your head steady while your partner gently pours water into the can.

Describe your observations.

Sketch ray diagrams to illustrate how you were able to see the coin after the water was added, but not before.

In drawing a ray diagram, it is often necessary to show the location of an observer. In

this module, we use the following symbol to show the observer: $\triangleright$.

Experiment 6.2

A. A pin is held vertically at the back of a square clear container of water, as shown below. The portion of the pin below the surface of the water is not shown in the side view diagram.

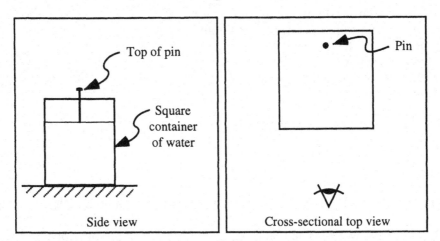

On a separate piece of paper, make an enlargement of the top view diagram.

On the enlargement, sketch rays to *predict* where the bottom of the pin would *appear* to be located to the observer shown. For simplicity, assume that light passes directly from water to air.

> *Note:* Do not try to draw a ray diagram in which the rays bend by exactly the correct amount. However, your diagram should be qualitatively correct.

On the basis of your ray diagram, would the bottom of the pin appear to be located closer, farther, or the same distance from the observer as the top of the pin? Explain your reasoning.

B. Set up the equipment and use the method of parallax to check your predictions.

Does your observation agree with your prediction? If not, how can you resolve the inconsistencies? Sketch a new ray diagram that is consistent with your observations.

C. Check whether the assumption that light from the pin passes directly from water to air, is reasonable. In other words, we would like to see whether the plastic has a negligible effect on the path taken by the light.

Devise and perform an experiment that would allow you to test whether this assumption is valid.

D. Imagine that you were to look at the pin from the location shown at right.

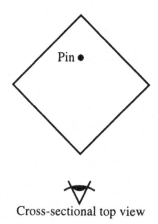

Cross-sectional top view

Sketch a ray diagram to illustrate what you would see. (*Hint:* What would you see if the left side or the right side of the container were covered?)

Check your predictions. If your observations are in conflict with your predictions, try to resolve the inconsistencies. Sketch a new ray diagram to account for your observations.

✔ Discuss your results with a staff member.

As you have observed, an object in water may appear to be in a different location to an observer out of the water. The place where the bottom of the pin appeared to be located is called the location of the image of the pin, or the *image location*.

Experiment 6.3

A. Suppose a nail were held vertically at the back of a circular beaker of
    water, as shown below. (The portion of the nail below the surface of
    the water is not shown in the side view diagram.)

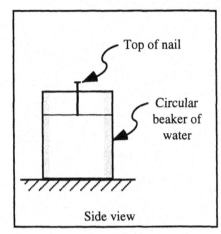

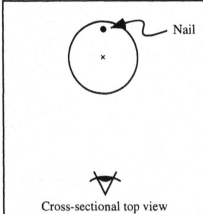

On a separate piece of paper, make an enlargement of the top view
diagram. Use a compass to draw an accurate circle.

Predict whether the image of the bottom of the nail is closer, farther,
or the same distance from the observer as the top of the nail. Sketch
a qualitatively correct ray diagram to support your prediction.
Explain. (*Hint:* Treat the nail as a point source of light.)

Use parallax to place a second nail at the location of the image of the
bottom of the nail, and check your predictions. If your prediction was
incorrect, find your error.

B. Three students are discussing their results from part A:

Student 1: *"I think the image is closer to me than the nail itself. The
closer something is, the bigger it looks. Because the image
of the nail appears larger than the nail itself, the image
must be closer to me than the nail."*

Student 2: *"But when I used parallax to determine the location of the
image of the nail, I found the image to be farther from me
than the nail."*

Student 3: *"That doesn't make sense. If the image were farther from
me than the nail, then how could I see the image? Wouldn't
the nail block my view of the image?"*

Do you agree or disagree with each of these students? Discuss your
reasoning with your partners.

McDermott & P.E.G., U.Wash./*Physics by Inquiry*

C. Consider two other locations of the nail: at the center of the beaker and near the front of the beaker, as shown below.

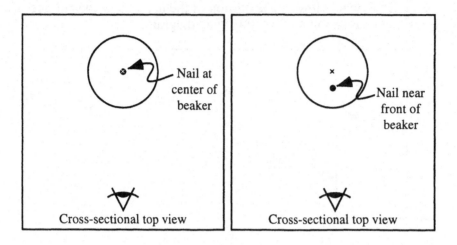

For each of these cases:

(1) Predict whether the image of the bottom of the nail would appear closer to the observer, farther from the observer, or the same distance from the observer as the top of the nail.

(2) Sketch a qualitatively correct ray diagram that supports your prediction.

Check your predictions, and resolve any inconsistencies.

D. Describe how the location of the image of the nail changes as the nail is moved from the far side of the beaker to the near side of the beaker.

✔ Discuss your results with a staff member.

Section 7. Image formation by convex lenses

Exercise 7.1

A. At right is a diagram of a light ray incident on a thin rectangular plate of glass. The line *AA'* is perpendicular to the plate.

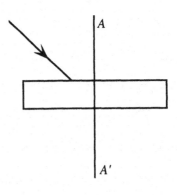

　　Sketch the continuation of the ray through the glass and into the air.

　　　How does the angle between the emergent ray and *AA'* compare to that between the incident ray and *AA'*? Explain.

　　　How would your answer change if the plate were thicker or thinner than the one shown above? Explain.

B. At right is a diagram of a light ray incident on a piece of glass with one spherical surface and one plane surface.

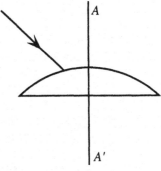

　　　How does the angle between the emergent ray and *AA'* compare to that between the incident ray and *AA'* ? Explain.

C. Suppose that the spherical surface in the diagram in part B were replaced by a spherical surface with a smaller radius of curvature as shown below. The angle between the incident ray and *AA'* is the same as in part B.

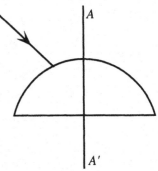

　　　How would the angle between the emergent ray and *AA'* compare to the corresponding angle in part B? Explain.

D. Suppose that the glass in part B were made of a type of glass with a greater index of refraction.

　　　How would the angle between the emergent ray and *AA'* compare to the corresponding angle in part B? Explain.

E. What factors determine the ability of a piece of transparent material, such as glass, to change the direction of an incident ray? Explain your reasoning.

　　　　McDermott & P.E.G., U.Wash./ *Physics by Inquiry*

A *lens* is an optical device that refracts light. The pieces of glass in the previous experiment are examples. Lenses can be divided into two categories according to shape: *convex* and *concave*. Convex lenses have at least one surface that curves outward and are thicker in the middle than at the edges. In contrast, concave lenses have at least one surface that curves inward and are thinner in the middle than at the edges.

The lenses in parts B and C in Exercise 7.1 are called *plano-convex* lenses. Another common type of convex lens has both surfaces curved outward. These are called *double-convex* lenses. In this section, we investigate image formation by convex lenses in air.

Experiment 7.2

Obtain two convex lenses of different diameters.

A. Look at various objects through the lenses. Describe what you see. What factors affect what you see? (Possibilities might include the following: distance from the object to the lens, distance of the lens to your eye, etc.) Record your observations.

B. Look at a coin through the lens with the larger diameter. Hold the coin so it is upright when you are looking at it without the lens.

Where can you hold the lens so that the coin appears to be:

(1) upside down?

(2) right side up?

(3) larger than it is in real life?

(4) smaller than it is in real life?

(5) the same size as it is in real life?

C. Hold the lens fixed in place about 30 cm from your eye. Place a coin behind the lens and very close to it. Slowly move the coin as far from the lens as you can. Describe your observations in words and with sketches.

Repeat the procedure above for the other lens. Do you notice any differences between the two lenses?

D. Hold the coin at arm's length in front of you. Place the lens directly in front of the coin, then move it slowly toward your eye. Keep the coin fixed in place as you move the lens.

Describe your observations in words and with sketches.

When we look at an object through a lens, we say that we are looking at an *image* of the object. In the preceding experiment, you found that the image can appear to be large or small. It can also appear to be right-side up *(erect)* or upside down *(inverted)*. In the following experiments, we examine in more detail the image formed by a lens.

Experiment 7.3

For this experiment, you will need the larger diameter lens from the preceding experiment.

A. Use a piece of clay to support the lens on a table as shown. Place your eye about 30 cm from the lens and look through the lens at a nail placed about a half meter beyond the lens.

Describe your observations of the image of the nail.

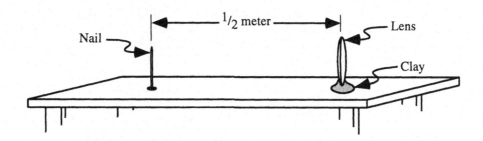

Does the location of the image seem to differ from the actual location of the nail? Explain.

Based on your observations, would you say that the image of the nail is closer to you, farther from you, or the same distance from you as the nail? Explain how you can tell from your observations.

B. Obtain a second nail that is identical to the first. Use the method of parallax to place the second nail at the location of the image of the first nail.

Does the image appear to be larger or smaller than the nail appears without the lens?

Is the location of the image consistent with your answer in part A? If not, how can you resolve the inconsistencies? (*Hint:* Can you determine the location of the image solely on the basis of the size of the image?)

Remove the second nail.

C. Place the first nail about 5 cm from the lens. Look at the nail through the lens with your eye about 30 cm from the lens. Use the method of parallax to determine the location of the image.

Is the image closer to you, farther from you, or the same distance from you as the nail?

Does the image appear to be larger or smaller than the nail appears without the lens? Can you use the apparent size of the image as a clue as to whether the image is closer to you than the nail?

D. Turn the lens around. Does the location of the image depend on which side of the lens is toward the nail?

✔ Discuss this experiment with a staff member.

Experiment 7.4

Obtain several convex lenses of different sizes and shapes.

A. Hold a lens in one hand so that light from a brightly lit distant source (e.g., an illuminated building) passes through the lens and falls on a sheet of paper held in your other hand. Move the paper back and forth until a clear image of the object appears on the paper.

Describe the appearance of the image. Is it erect or inverted? Is it larger or smaller than the object?

Turn the lens around. Describe what happens to the image.

Move the paper back and forth. Describe what happens to the image.

B. Repeat the process for each of the remaining lenses.

Describe the similarities and differences among the various lenses.

The line through the center of the lens and perpendicular to the plane of the lens is called the *principal axis*. (See the figure below.) The point of intersection of this axis and the image of a very distant object is called the *focal point*. The distance of the focal point from the center of the lens is called the *focal length*.

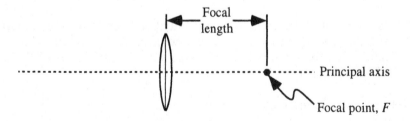

Experiment 7.5

For each lens that you used in the preceding experiment:

Is there a focal point on each side of the lens? If so, how does the focal length on one side of the lens compare with that on the other? Explain how you can tell from your observations.

It is customary to use a single symbol f to denote the focal length of a lens. Explain why two symbols, one for each focal point, are not necessary.

The image formed by the convex lens in the preceding experiment is called a *real image.* Such an image can be seen on a sheet of paper or a screen placed at the location of the image. A real image is different from the type of image formed by a plane mirror, which is called a *virtual image.* A virtual image cannot be seen on a screen placed at the location of the image.

Experiment 7.6

Place two meter sticks end-to-end on a table so that the zero markings are at the same point. Place the center of a lens above this point. Use a piece of clay as shown.

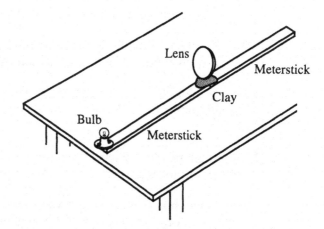

A. Find the focal points on each side of the lens and mark each of these points.

Describe the method you used to find the focal point.

B. Place a lighted bulb at the far end of one of the meter sticks.

Locate the image of the bulb by parallax. Also locate the image by using a piece of paper as a screen.

Compare the results of these two methods.

Describe the image and record your observations.

(1) Is it real or virtual?

(2) Is it erect or inverted?

(3) Does it appear to you to be larger, smaller, or the same size as the object?

(4) Is it on the same side of the lens as the object or on the opposite side?

(5) Note where the image is located relative to the lens and its focal points.

C. Repeat part B for each of following locations for the lighted bulb:

(1) at a distance between 1 m and $2f$ from the lens

(2) at a distance $2f$ from the lens

(3) at a distance between $2f$ and f from the lens

(4) at a distance f from the lens

(5) at a distance between f and $f/2$ from the lens

(6) at a distance $f/2$ from the lens

(7) at a distance less than $f/2$ from the lens

Do the two methods for determining the location of the image always work? If not, note which method fails and when it fails.

Place the lighted bulb as close to the lens as possible. Describe the image in this case. Record your observations.

D. Summarize your observations in words. If necessary, make additional observations to answer the following questions:

As an object is moved toward a lens from a large distance away, how does the location of the image vary?

As the object is moved toward the lens, how does the apparent size of the image vary?

For what range of object locations is the image erect? For what range of object locations is the image inverted?

Where must the object be located in order for the image to be on the same side of the lens as the object? Where must the object be located in order for the image to be on the opposite side of the lens as the object?

✔ Discuss this experiment with a staff member.

Section 8. Image formation and ray diagrams

In this section, we develop an algorithm, or a set of rules, for predicting the position and appearance of the image formed by a convex lens. This procedure, which is based on the tracing of light rays through a lens, is a powerful tool that we will apply and extend throughout our study of geometrical optics.

Exercise 8.1

The diagram below shows a nail near a convex lens. The image formed by the lens is shown on the diagram.

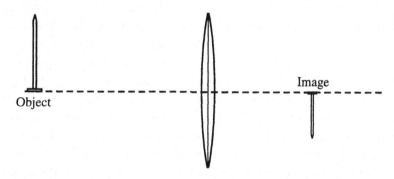

A. From the tip of the nail, draw several rays, some of which pass through the lens.

Show by a sketch how you can use these rays to account for the formation of the image of the tip of the nail. Explain.

B. Draw similar sketches for two other points on the nail.

C. Explain how you decided to draw the rays in the way that you did in parts A and B.

D. A convex lens is often called a *converging lens*.

Is this term appropriate? Explain why or why not.

Exercise 8.2

A lens is held at a very great distance from a tall building so that a clear image of the building is formed on a piece of paper. Consider two points on the building, *P* and *Q*. *P* is on the principal axis, *Q* is directly above it at the top of the building.

A. Draw a diagram that shows several rays from point *P* that reach the lens.

   How are these rays oriented with respect to one another and the principal axis? Explain.

   Where do these rays converge? Explain your reasoning.

B. On the diagram above, sketch some rays from point *Q* that reach the lens.

   How are these rays oriented with respect to one another? Explain.

   Where do these rays converge? Explain your reasoning.

The image of the building in the preceding exercise lies in the *focal plane* of the lens. This plane intersects the principal axis at the focal point and is perpendicular to the principal axis.

Experiment 8.3

A very small object is placed at the focal point of a convex lens. Where would you expect the image to be located?

Draw a diagram to illustrate your prediction. (*Hint:* Consider the effect of reversing the direction of light through a lens or a beaker of water. You did such an experiment in an earlier section on refraction.)

Set up the apparatus and check your prediction.

Exercise 8.4

Draw an enlarged side view diagram of a convex lens.

A. Draw a ray incident on the lens that passes through the center of the lens. The ray should be incident on the lens at an angle with the principal axis.

   How does the direction of the incident ray compare to the direction of the ray that emerges from the lens? Explain your reasoning.

B. Suppose that the lens that you drew were made much thinner. Use a diagram to justify the following statement:

   The path of any ray passing through the center of a thin lens is essentially undeviated.

✔ Discuss Experiment 8.1 through Exercise 8.4 with a staff member.

The location of the image for any object can be found through the use of a ray diagram. Analysis of the preceding experiments and exercises suggests a method for tracing the path of light rays from a point on an object, through a lens, to a corresponding image point. Below, we apply this technique to the construction and interpretation of the image formed by a convex lens.

Exercise 8.5

In drawing a ray diagram for a lens, we begin by showing the lens, the principal axis, and the locations of the two focal points. A simple sketch of the object is then drawn at the proper distance from the center of the lens, usually on the left side of the lens.

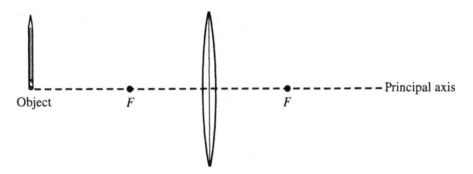

As we found in Exercise 8.1, we can think of all the rays from a single point on the object that pass through the lens as converging at a single corresponding image point.

There are three special rays from each object point for which it is particularly easy to determine the path through the lens. These are shown in the following sketch. Copy this diagram in your notebook.

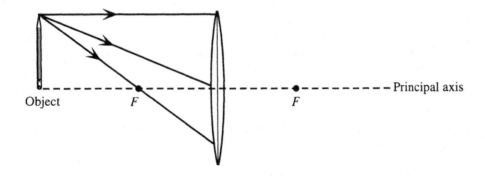

A. Do you expect these three rays to intersect after they pass through the lens? Explain.

Draw the continuation of each of these rays on the other side of the lens. In each case, explain your reasoning.

How would you interpret the intersection of these three rays?

These three rays, which are called *principal rays,* are only a few of the infinitely many that we might draw from one point on the object.

B. Choose another ray from the same point on the object. Continue that ray through the lens and out the other side. Explain how you were able to determine the direction of that ray after it passed through the lens.

In drawing a ray diagram for a thin lens, it is customary to treat rays either as being refracted all at once at the center of the lens or to treat the lens as a line with no thickness. Refraction actually takes place at the two surfaces.

C. Choose another point on the object and use the three principal rays from that point to determine the location of the corresponding point on the image.

D. Consider a point on the object that lies on the principal axis. Where is the image of this point located? Explain.

E. Sketch the entire image at the appropriate location on your diagram.

✔ Discuss this exercise with a staff member.

Exercise 8.6

A nail is placed near a lens. The top of the nail is above the top of the lens, as shown in the diagram below.

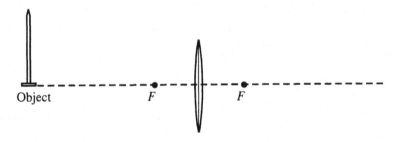

A. Will the lens form a complete image of the nail? Sketch a ray diagram to illustrate your answer.

B. How many of the principal rays must you draw to be able to locate an image? Explain.

McDermott & P.E.G., U.Wash./*Physics by Inquiry*

Exercise 8.7

A pencil is placed near a converging lens of focal length f. Three observers are located as shown.

A. Draw a ray diagram in which you use the principal rays to determine the location of the image of the pencil.

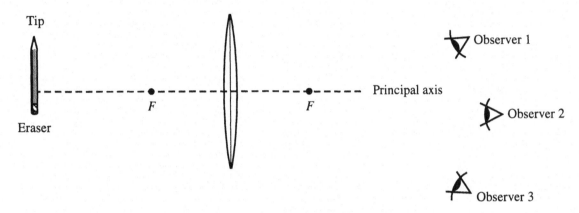

B. Which observer(s) can see the image of the pencil tip? Explain.

C. Which observer(s) can see the image of the eraser? Explain.

D. Which observer(s) can see the entire image of the pencil? Explain.

E. Suppose that you were to place your eye at the location of an observer who can see the entire image of the pencil.

Which would *appear* to be larger to the observer: the image of the pencil (with the lens in place) or the pencil (with the lens removed)? Explain how you can tell from the ray diagram.

If you were to use a ruler to measure the length of the pencil and the length of the image, which would be larger? Explain how you can tell from the ray diagram.

✔ Discuss your reasoning with a staff member.

McDermott & P.E.G., U.Wash./*Physics by Inquiry*

Exercise 8.8

In Exercise 8.5, we developed a procedure for drawing ray diagrams for situations in which a convex lens forms a real image. In this exercise, we extend this procedure to include the case that the image is virtual.

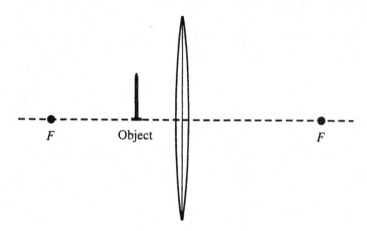

A. Copy this diagram in your notebook and draw the three principal rays from the tip of the nail to the lens.

Do these rays intersect on the right side of the lens?

Would an observer on the right side of the lens see an image? If so, where is the image located? (*Hint:* Consider your observations using convex lenses in Section 7.)

Explain how you would find the location of the image by using the principal rays. Recall that it is customary to use dashed lines to indicate the extension of a ray into a region in which the light does not pass.

B. Is the image erect or inverted? Is the image larger or smaller than the object? Explain how you can tell from the ray diagram.

Exercise 8.9

An object is placed near a lens along the principal axis.

A. For each of the following distances from the object to the lens, draw a ray diagram to determine the location of the image:

(1) greater than $2f$ (4) at f (7) less than $f/2$

(2) at $2f$ (5) between f and $f/2$

(3) between $2f$ and f (6) at $f/2$

B. For each case in part A:

Is the image erect or inverted?

Is the image real or virtual?

How does the size of the image compare to the size of the object?

Which would *appear* to be larger to an observer: the image of the object (with the lens in place) or the object (with the lens removed)? Explain how you can tell.

C. Compare your results for the image locations in part A with your observations in Experiment 7.6. Are they consistent? If not, resolve any discrepancies.

Experiment 8.10

In this experiment, be sure to make all predictions before performing the experiments. Draw ray diagrams to justify your predictions.

A lens, a bulb, and a screen are arranged as illustrated in the following diagram. The screen is at the location of the image of the bulb.

Bulb

 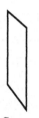

Lens and clay   Screen

A. Predict what would happen to the image if the top half of the lens were covered by a mask. Explain your reasoning.

Does it matter on which side of the lens the mask is placed?

B. Predict what would happen to the image if a mask with a small hole in the center (smaller than the object) were placed in front of the lens.

How would your answer differ if:

- the hole were made larger than the object

- the hole were moved so it is not at the center of the lens

C. Perform each of the above experiments and check your predictions. If your predictions were incorrect, draw new ray diagrams that are consistent with what you observe.

Section 9. Image formation and the thin lens equation

In the preceding section, we introduced a diagrammatic representation that enables us to make predictions about the image formed by a thin lens. In this section, we develop an algebraic representation for the relationships among the distance of an object from a lens, the distance of an image from a lens, and the focal length of the lens.

In the algebraic analysis of image formation by a single lens, we use the following symbols.

> The symbol s represents the object distance measured from the center of the lens along the principal axis and the symbol s' represents the corresponding image distance from the center of the lens. The symbols F and f designate the focal point and focal length on the same side of the lens as the object, and the symbols F' and f' designate the focal point and focal length on the other side of the lens.

In developing an algebraic equation that relates the object distance and image distance, we need to establish an algebraic sign convention. We adopt the following convention.

> *Algebraic Sign Convention for object and image distances for a lens*
>
> The object distance, s, is considered positive. (The object is usually placed to the left of the lens.) The image distance, s', is considered positive when the image is on the opposite side of the lens from the object and negative when it is on the same side of the lens as the object. The focal length of a convex lens is considered positive.

Experiment 9.1

> In this experiment, we develop a quantitative relationship between the object distance and the image distance for a convex lens. We repeat the procedure of Experiment 7.6, in which we made qualitative observations of how the distance of an object from a lens affects the image.
>
> A. Place an object at a great distance from a convex lens. Record the values of s and s' in a table in your notebook.
>
> Move the object closer to the lens in several stages. (You may use the same locations for the object as you did in the Experiment 7.5.) For each location of the object, record the values of s and s'.
>
> Include an entry in your table for an image located at F'. What are the corresponding values of s and s'?

B. A convenient way to graph your results is to use new variables, x and x'. Let $x = s - f$, and $x' = s' - f'$.

> What physical interpretation can you give to the variables x and x'?

> Where is the object when x is positive, when x is negative and when x is zero? Where is the image when x' is positive, when x is negative, and when x is zero?

(1) Add two columns to your table, one for x and one for x'. Enter values for x and x' corresponding to each value of s and s' in your table.

(2) Graph x as a function of x'.

(3) As your graph indicates, the relationship between x and x' is clearly not linear. You might guess from your study of mathematics that the resultant curve is a hyperbola.

> To test whether the curve is a hyperbola, plot a graph of x as a function of $1/x'$.

> > Is the relationship between x and $1/x'$ linear? Explain how you can tell from your graph.

> > Find the slope of the graph and call it κ.

> > Write an equation relating x, x', and κ.

> > The constant κ is related to the focal length, f. Make an educated guess as to this relationship, by considering the units of κ. Check your guess.

> > Write an equation relating x, x', and f.

✔ Check your reasoning with a staff member.

The equation that you derived in the preceding experiment is known as the Newtonian form of the thin lens equation. This form of the thin lens equation is seldom used. The Gaussian form, which is given in the following exercise, is more useful in practice.

Exercise 9.2

The Gaussian form of the thin lens equation is:

$$\frac{1}{s} + \frac{1}{s'} = \frac{1}{f}$$

A. Derive the Gaussian form of the thin lens equation from the Newtonian form.

B. Use your measurements from the preceding experiment to plot $1/s$ as a function of $1/s'$.

   What is the shape of your graph?

   What feature of the graph can you use to determine the focal length of the lens? Explain your reasoning.

The expressions that we have derived that relate object and image distances to the focal length of a lens are very useful for making predictions and designing simple optical instruments. It should be noted, however, that these equations apply only to thin lenses, in which the thickness of the lens is small compared to the other distances involved.

Exercise 9.3

A. Use the thin lens equation to find the image location, s, for an object placed at the following values of s: $s = 4f,\ 2f,\ 3f/2,\ f,\ 3f/4,\ f/2,\ f/4$. Show your work in each case.

B. Not all of the values of s in part A result in a positive value for s'.

   How would you interpret negative value for s'. How would you interpret the failure of the equation to yield a number for s'? (*Hint:* Draw on your experience in working with lenses in the laboratory and on your experience in drawing ray diagrams.)

Exercise 9.4

A. We now have three ways of finding the location of an image formed by a thin convex lens: by using parallax, by using a screen, and by using an equation.

   Discuss which of the methods work for finding the image locations for each of the images in the preceding exercise. You have used the first two of these methods in Experiment 7.6.

If one of the methods fails for some of the object distances, discuss why that method fails.

B. Give an operational definition for a real image and for a virtual image. (You may need to refer to the discussion of operational definitions in *Properties of Matter* in Volume I.)

✔ Check your reasoning with a staff member.

Experiment 9.5

In this experiment, you will make several predictions. Record your predictions before performing any experiments. Explain your reasoning.

Suppose that a lighted bulb is placed about 2 meters from a convex lens. Imagine that you are looking at the bulb through the lens with your eye about 30 cm from the lens.

A. Predict what you would observe. Where would the image be located?

Predict what you would observe if a piece of plain white paper were placed at the location of the image.

Predict what you would observe if the paper were placed closer to the lens.

Predict what you would observe if the paper were moved to a location farther from the lens.

Perform the above experiments and record your observations. Try to resolve any discrepancies between your predictions and your observations.

Are your predictions and your observations consistent with the implications of the thin lens equation? Explain.

B. Predict what you would observe on a sheet of paper at the image location if the lens were removed.

How would what you would observe differ from what you observe when the lens is present?

With the lens still removed, predict what will happen as the piece of paper is moved toward and away from the bulb. Explain your reasoning.

Perform the above experiments and record your observations. Try to resolve any discrepancies between your predictions and your observations.

Exercise 9.6

As illustrated below, the size of the image formed by a convex lens depends on the size and location of the object. We define the magnification, *m*, as the ratio of the height of the image, *y′*, to the height of the object, *y*.

$$m = \frac{y'}{y}$$

A. Find the relative size of the image with respect to the object in terms of the image and object distances, *s′* and *s*. (*Hint:* Look for similar triangles in the diagram above.) Explain.

B. By convention, the magnification *m* of an inverted image is chosen to be negative. Note that this implies that *y′* is negative for inverted images, such as the real image above.

   Derive an expression for *m* in terms of *s′* and *s* that is consistent with this convention.

Experiment 9.7

Consider a convex lens of focal length *f*. Suppose that an object is placed at a distance less than *f* from the lens.

A. Is the magnification *m* positive or negative? Explain your reasoning.

B. Is the absolute value of the magnification greater than, equal to, or less than 1? Explain your reasoning.

C. Suppose that an observer who was originally located to the right of the image above moved further away from the lens.

   Would the magnification of the lens change? Explain.

   Would the size of the image as seen by that observer change? Would the size of the object as seen by that observer (with the lens removed) change? Explain.

   ✔ Check your results with a staff member.

Section 10. Image formation by concave lenses

In this section, we examine the behavior of concave lenses. Concave lenses have at least

one surface that curves inward and are thinner in the middle than at the edge.

Experiment 10.1

Obtain two concave lenses of different diameters.

A. Look at various objects through the lenses. Describe what you see.
What factors affect what you see? (Possibilities might include the
following: distance from the object to the lens, distance of the lens to
your eye, etc.) Record your observations.

B. Look at a coin through the lens with the larger diameter. Hold the coin
so it is upright when you are looking at it without the lens.

Where, if at all, can you hold the lens so that the coin appears to be:

(1) upright?

(2) inverted?

(3) smaller than it is in real life?

(4) larger than it is in real life?

(5) the same size as it is in real life?

C. Hold the lens fixed in place about 20 cm from your eye. Place a coin
behind the lens and very close to it. Slowly move the coin as far from
the lens as you can. Describe your observations in words and with
sketches.

Repeat the procedure above for the smaller lens. Do you notice any
differences between the two lenses?

D. Hold the coin at arm's length in front of you. Place the lens directly in
front of the coin, then move it slowly toward your eye. Keep the coin
fixed in place as you move the lens.

Describe your observations in words and with sketches.

Experiment 10.2

For this experiment, you will need the larger diameter lens from the
preceding experiment.

A. Use a piece of clay to support the lens on a table as shown in the
following diagram. Place your eye about 10 cm from the lens and look
through the lens at a nail placed about 40 cm beyond the lens.

McDermott & P.E.G., U.Wash./*Physics by Inquiry*

Describe your observations of the image of the nail.

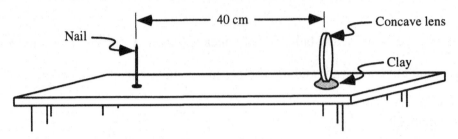

Does the location of the image seem to differ from the actual location of the nail? Explain.

Based on your observations, would you say that the image of the nail is closer to you, farther from you, or the same distance from you as the nail? Explain how you can tell from your observations.

B. Obtain a second nail, identical to the first. Use the method of parallax to place the second nail at the location of the image of the first nail.

Does the image appear to be larger or smaller than the nail appears without the lens?

Is the image location consistent with your answer in part A? If not, how can you resolve the inconsistencies? (*Hint:* Can you determine the image location solely on the basis of the size of the image?)

Remove the second nail.

C. Place the first nail close to (but not touching) the lens. Look at the nail through the lens with your eye about 10 cm from the lens. Use the method of parallax to place the second nail at the location of the image.

Is the image closer to you, farther from you, or the same distance from you as the nail?

Does the image appear to be larger or smaller than the nail appears without the lens?

D. Reverse the lens. Does the location of the image depend on which side of the lens is toward the nail?

Experiment 10.3

Compare the behavior of concave and convex lenses. In particular:

*   Is it possible to form both real and virtual images using either type of lens?

*   Is it possible to form both erect and inverted images using either type of lens?

✔ Discuss Experiments 10.1 through Exercise 10.3 with a staff member.

Experiment 10.4

Obtain several concave lenses of different sizes and shapes.

A. Hold a lens in one hand so that light from a brightly lit source (e.g., an illuminated building) passes through the lens and falls on a sheet of paper held in your other hand. Try to place the paper at the image location.

Is it possible for an image of the distant object to appear on the paper? Explain how you can tell from your observations.

Would your answer above be different if you reversed the lens?

B. Repeat the process for each of the remaining lenses. Describe the similarities and differences of your results with various lenses.

C. On the basis of your observations, is the image formed by a concave lens of a distant object real or virtual?

In a previous section, we developed a method for determining the location of the focal point and focal length of a convex lens. The focal point is the point of intersection of the principal axis and the image of a very distant object. In the following experiment, we determine the focal length of a concave lens.

Experiment 10.5

Find the focal length for both sides of each lens from Experiment 10.4.

Is there a focal point on each side of the lens? If so, how does the focal length on one side of the lens compare with that on the other? Explain how you can tell from your observations.

It is customary to use a single symbol f to denote the focal length of a lens. Explain why two symbols, one for each focal point, are not necessary.

✔ Discuss Experiments 10.3 and 10.4 with a staff member.

For a convex lens, the image of a distant object is on the opposite side of the lens from the object. As you have seen, for a concave lens, the distant object and its image are both on the same side of the lens. The following sign convention is consistent with our observations and also with the sign convention for s and s'.

     McDermott & P.E.G., U.Wash./ *Physics by Inquiry*

Algebraic sign convention for the focal length of a lens

When a distant object and its image are on opposite sides of the lens, the focal length is considered positive. (e.g., a convex lens)

When a distant object and its image are on the same side of the lens, the focal length is considered negative. (e.g., a concave lens)

Exercise 10.6

The diagram below shows a nail near a concave lens. The image formed by the lens is also shown.

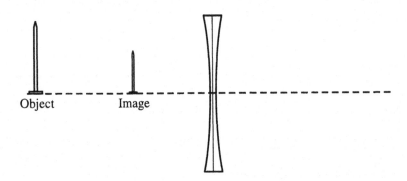

A. From the tip of the nail, draw several rays. Some of the rays that you have drawn should pass through the lens.

Show with your sketch how you can use these rays to account for the formation of the image of the tip of the nail. Explain.

B. Draw similar sketches for two other points on the nail.

C. Explain how you decided to draw the rays in the way that you did in parts A and B.

D. A concave lens is often called a *diverging lens*.

Is this term appropriate? Explain why or why not.

You have already used ray diagrams to determine the location of an image formed by a convex lens. Below, we apply this technique to the construction and interpretation of the image formed by a thin concave lens. Recall that a lens may be regarded as a thin lens if its thickness is small compared to the other distances involved.

Exercise 10.7

The diagram below shows a thin concave lens, the principal axis of the lens, an object, and the location of the two focal points.

As in Exercise 8.1, we can think of all the rays from a single point on the object that pass through the lens as appearing to originate from a single corresponding image point.

Copy the diagram below in your notebook.

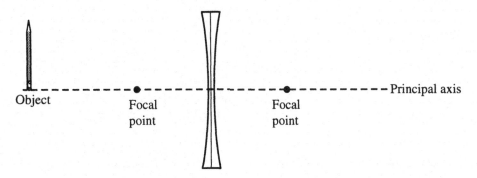

A. On your diagram trace a ray parallel to the principal axis that extends from the tip of the object to the lens.

How would you draw the continuation of this ray through the lens? Is your answer consistent with the idea that this is a diverging lens?

B. Trace another ray from the tip of the object that is directed toward the center of the lens.

Does this ray change direction as it passes through the lens, assuming the lens is thin? Explain.

C. Determine where the image of the tip of the object is located. Explain your reasoning.

Do the rays that you have drawn actually intersect at a point, or do they only appear to have passed through the same point? Explain.

D. Trace a third ray from the tip of the object that is initially directed toward the focal point on the far side of the lens.

Continue the ray through the lens and out the other side. Explain how you determined the direction of the ray on the other side.

How does the direction of the continuation of this ray compare to the orientation of the principal axis?

The three rays that you have drawn are called *principal rays*. Although tracing principal rays is a convenient tool for determining the location of an image formed by a lens, these rays are only a few of the infinitely many that we might draw from one point on the object.

E. Trace a fourth ray from the tip of the object to the lens. Continue this ray through the lens and out the other side of the lens. Explain how you determined the direction in which this ray continues on the other side.

✔ Discuss your reasoning with a staff member.

Exercise 10.8

For this exercise you will need your ray diagram from the preceding exercise.

A. Choose a point on the object other than the tip of the object. On the ray diagram, trace rays from this point to determine where the image of this point is located. (You may want to use ink of a different color.) Explain the reasoning you used in determining your answer.

B. Consider a point on the object that lies on the principal axis. Trace rays to show where the image of this point is located.

C. Sketch the entire image of the object at the appropriate location on your diagram.

✔ Discuss your completed ray diagram with a staff member.

Exercise 10.9

A nail is placed near a converging lens as shown. The nail is taller than the lens.

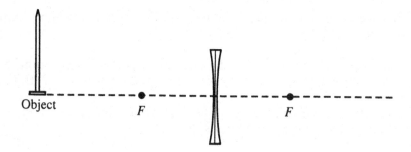

A. Will the lens form a complete image of the pencil? Sketch a ray diagram to illustrate your answer.

B. How many of the principal rays must you draw to be able to locate an image? Explain.

✔ Check your answers with a staff member.

We have used ray tracing techniques to predict the location of images formed by lenses. In the following experiments, we develop an alternative method.

Experiment 10.10

Place two meter sticks end-to-end on a table so that the zero markings are at the same point. Place the center of a concave lens above this point. Use a piece of clay as shown.

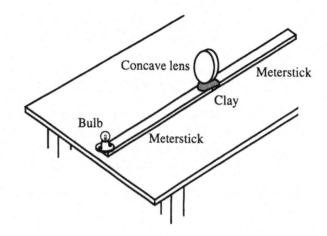

A. Find and mark the focal points of the lens.

B. Place a lighted bulb at the far end of one of the meter sticks.

Locate the image of the bulb by parallax.

Is it also possible to locate the image by using a piece of paper as a screen? Explain your reasoning.

Describe the image and record your observations.

(1) Is it real or virtual?

(2) Is it erect or inverted?

(3) Does it appear to you to be larger, smaller, or the same size as the object?

(4) Is it on the same side of the lens as the object or on the opposite side?

(5) Note where the image is located relative to the lens and its focal points.

C. Repeat part B for each of following locations for the lighted bulb:

(1) at a distance between 1 m and $2f$ from the lens

(2) at a distance $2f$ from the lens

(3) at a distance between $2f$ and f from the lens

(4) at a distance f from the lens

(5) at a distance between f and $\frac{f}{2}$ from the lens

(6) at a distance $\frac{f}{2}$ from the lens

(7) at a distance less than $\frac{f}{2}$ from the lens

Place the object as close to the lens as possible. Describe the image in this case. Record your observations.

D. Summarize your observations in words. If necessary, make additional observations to answer the following questions:

As an object is moved toward a concave lens from a large distance away, how does the location of the image change?

As the object is moved toward the concave lens, how does the apparent size of the image change?

For what range of object locations, if any, is the image erect? For what range of object locations, if any, is the image inverted?

Where, if anywhere, must the object be located in order for the image to be on the same side of the lens as the object? Where, if anywhere, must the object be located in order for the image to be on the opposite side of the lens as the object?

✔ Discuss your results with a staff member.

Exercise 10.11

An object is placed near a concave lens along the principal axis.

A. For each of the following distances from the object to the lens, draw a ray diagram to determine the location of the image:

(1) greater than $2f$ (5) between f and $f/2$

(2) at $2f$ (6) at $f/2$

(3) between $2f$ and f (7) less than $f/2$

(4) at f

B. Compare your results for the image locations in part A with your observations in part C of Experiment 10.10. Are they consistent? If not, resolve any discrepancies.

C. For each case in part A:

Is the image erect or inverted?

How does the size of the image compare to the size of the object?

Which would *appear* to be larger to the observer: the image of the object (with the lens in place) or the object (with the lens removed)? Explain how you can tell.

In the following experiment, we develop an algebraic representation for the relationships among the object distance, the image distance and the focal length of a thin concave lens.

Experiment 10.12

Arrange two meter sticks end-to-end on a table in the same way as you did for Experiment 10.10. Place the center of a concave lens with a known focal length above the junction of the two meter sticks. Use clay to hold the lens upright.

In this experiment be sure to indicate the proper algebraic signs for s, s' and f. Refer to the sign conventions at the beginning of Section 9 and earlier in this section.

A. Place a lighted bulb at the far end of one of the metersticks. Measure the distance of the bulb from the lens.

   Determine the corresponding image distance s'.

B. Move the object closer to the lens in small steps. For each location of the object, record the values of s and s' in a table.

   Include an entry in your table for the location of the image of a very distant object. What is the value of s' in this case?

C. Use your measurements to plot $1/s$ as a function of $1/s'$. Describe the shape of the resulting graph.

   Does the graph have a single value for the slope? If so, determine the slope of the graph.

   At which point does your graph intercept the vertical axis? How is the vertical intercept related to the focal length of the lens? Explain.

   Is the algebraic sign that you determined for f from your graph consistent with the sign convention discussed in the text after Experiment 10.5?

D. Write an equation relating s, s', and f.

   Compare the equation that you obtain for diverging lenses with the thin lens equation that you derived for converging lenses.

   Can the same equation be used for both types of lenses? As part of your answer, discuss the critical importance of our conventions for the algebraic signs of s, s', and f.

✔ Discuss your results with a staff member.

Experiment 10.13

Use the thin lens equation to find the image distance, s', for an object located (1) very far from the lens, (2) at a distance $2f$ from the lens, (3) at a distance f from the lens, and (4) at a distance less than f from the lens.

   Do your answers agree with your observations? If not, try to resolve any inconsistencies.

McDermott & P.E.G., U.Wash./ *Physics by Inquiry*

Section 11. Image formation by curved mirrors

Curved mirrors can be divided into two categories according to shape: convex and concave. The mirrored surface of a concave mirror curves inward and that of a convex mirror curves outward. As with a lens, we define the *principal axis* of a mirror as the line of symmetry for the curved surface. The point at which this line intersects the mirror is called the *vertex* of the mirror. The object distance, image distance, and focal length are measured from the vertex of the mirror.

In this section, we investigate image formation by curved mirrors, beginning with spherical concave mirrors.

Concave mirrors

Experiment 11.1

Obtain two concave mirrors of different diameters.

A. Look at various objects in the mirrors. Record your observations.

What factors affect what you see?

B. Hold a coin upright in front of the larger diameter mirror. Look at the image.

Where can you hold the mirrors so that the image is:

(1) upright?

(2) upside down?

(3) larger than it is in real life?

(4) smaller than it is in real life?

(5) the same size as it is in real life?

C. Hold the larger diameter concave mirror fixed in place at arm's length in front of you. Place a coin in front of the mirror and very close to it. Slowly move the coin toward your eye. Describe your observations in words and with sketches.

D. Does a concave mirror have a focal point? If so, are there one or two focal points? Explain how you can tell.

> For your larger concave mirror, measure the distance from the vertex to the focal point(s).

Is the image of a distant object formed by a concave mirror real or virtual? Explain how you can tell.

Is it possible to form both real and virtual images using a concave mirror?

Is it possible to form both erect and inverted images using a concave mirror?

Experiment 11.2

The diagram below shows an object held near a concave mirror. The image formed by the mirror is shown.

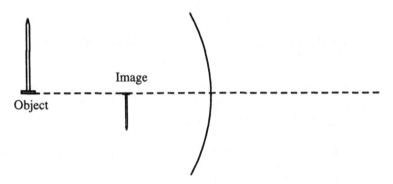

A. From the tip of the nail, draw several rays. Some of the rays that you have drawn should intersect the mirror.

> Show with your sketch how you can use these rays to account for the formation of the image of the tip of the nail. Explain.

B. Draw similar sketches for two other points on the nail.

C. Explain how you decided to draw the rays in the way that you did in parts A and B.

          McDermott & P.E.G., U.Wash./*Physics by Inquiry*

Exercise 11.3

For a mirror, we can identify three principal rays that are analogous to those for a lens. These rays are shown below for the tip of a nail.

Copy the diagram in your notebook.

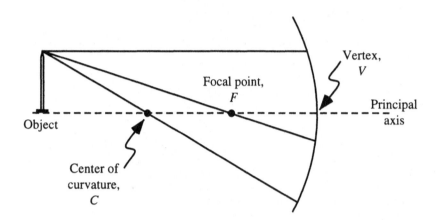

A. Do you expect these three rays to intersect after they are reflected from the mirror? Explain.

> Draw the continuation of each of these rays. Base your drawing on your knowledge of what you see when you look in a concave mirror. Explain your reasoning.

> How would you interpret the intersection of these three rays?

B. Draw the ray from the tip of the nail to the vertex *(V)* and show the reflected ray.

> You could have determined the direction of the reflected ray in two ways. Discuss both briefly.

C. How many of the four rays that you drew above are necessary in order to determine the image location? Explain.

D. Choose another point on the object and draw rays to determine the location of the corresponding point on the image.

E. Sketch the entire image at the appropriate location on your diagram.

✔ Discuss this exercise with a staff member.

Experiment 11.4

Obtain several concave mirrors of various radii of curvatures.

A. Perform an experiment that enables you to determine the location of the focal point for each mirror. Record the focal length and the radius of curvature for each mirror.

B. Plot a graph of focal length versus radius.

Determine the slope of the graph. Write an expression for the focal length in terms of the radius of curvature.

In working with lenses, we found that it was necessary to establish a sign convention in order to formulate an algebraic relationship between the object and image distance. The same is true for curved mirrors, however, the convention we adopt is different from that for lenses. The following is the convention that we use for mirrors in this module.

> *Algebraic sign convention for object and image distances for a curved mirror*
>
> The object distance, s, is considered positive. (The object is usually placed to the left of the mirror.) The image distance, s', is considered positive when the image is on the same side of the mirror as the object and negative when it is on the opposite side of the mirror as the object. The focal length of a concave mirror is considered positive.

Experiment 11.5

In Experiment 10.12, we found that the thin lens equation for convex lenses can also be applied to concave lenses. However, the proper choice of sign conventions is of critical importance.

In this experiment, we explore whether the same equation might also apply to a concave mirror by testing its validity at a few points.

A. Place a small object at the following locations:

- at the focal point of the mirror
- at the center of curvature of the mirror
- at a great distance from the mirror

In each case, determine the location of the image in terms of R, the radius of curvature of the mirror.

B. Check whether the object distances and image distances in part A are
consistent with the thin lens equation.

C. If a large curved mirror is available, check additional object locations
as well, such as $R = 3R/2$, $3R/4$, $R/2$, and $2R/3$.

Spherical aberration in concave mirrors

Experiment 11.6

Two rays of light from a distant source are
incident on a mirror that consists of one-half of
a sphere. The dashed line in the figure, which
shows the entire surface of the sphere, is given
for your convenience.

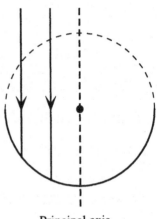

Principal axis

Draw an enlargement of the mirror on graph
paper. Use a compass to draw an accurate
semicircle.

A. Carefully draw the corresponding reflected
rays. Base your construction on your
knowledge of how light is reflected from a
plane mirror. Use a ruler and a protractor.

Do the reflected rays all pass through
the same point?

B. Suppose that all the rays from a given point on an object to do not
meet at a single point on the image. What effect would you expect this
to have on the appearance of the image? Explain.

Based on your observations in part A, would you expect a
spherical mirror to have a well defined focal point? Explain.

Suppose you were to paint over part of the mirror so that only light
incident on the mirror near the principal axis is reflected. Would
you expect the focal point to be better defined in this case?
Explain. Draw a diagram to support your answer.

Spherical aberration is the term given to the failure of a spherical mirror or lens to

form a sharp image of an object. For both, the effect is small for rays incident near the

principal axis. In spite of this inherent shortcoming, most mirrors and lenses are

spherical in shape since spherical surfaces are relatively easy to construct.

Experiment 11.7

A. Draw an accurate cross-section of a parabolic mirror on a piece of graph paper. Use your knowledge of the equation for a parabola as a guide.

Draw several rays to convince yourself that a parabolic mirror reflects all the rays parallel to the principal axis so that they pass through a single point, the focus of the mirror.

B. Imagine that you were to place a small lighted bulb at the focus of a spherical mirror and another bulb at the focus of a parabolic mirror.

Describe what would happen to the reflected light in both cases.

Convex mirrors

Experiment 11.8

A. Repeat Experiments 11.1 through 11.4 using convex mirrors instead of concave mirrors.

B. Compare the behavior of concave and convex mirrors. In particular:

- Is it possible to form both real and virtual images using either type of mirror?

- Is it possible to form both erect and inverted images using either type of mirror?

For a concave mirror, the image of a distant object is on the same side of the mirror as the object. For a convex mirror, a distant object and its image are on opposite sides of the mirror. The following sign convention is consistent with these observations and also with the sign convention for s and s' for curved mirrors.

Algebraic sign convention for the focal lengths of a curved mirror

When a distant object and its image are on opposite sides of the mirror, the focal length is considered positive. (e.g., a concave mirror)

When a distant object and the image are on the same side of the mirror, the focal length is considered negative. (e.g., a convex mirror)

Experiment 11.9

Explore whether the thin lens equation can be applied to convex mirrors. Perform the same test as in Experiment 11.5, using the same object locations as you did in that experiment.

Experiment 11.10

In this experiment, be sure to make all your predictions before performing the experiments. Draw ray diagrams to justify your predictions.

Place a screen next to a long filament bulb as shown in the diagram below. Place a concave mirror at the correct location to produce a sharp image of the bulb on the screen.

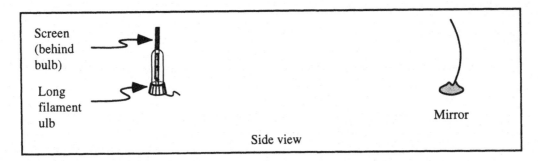

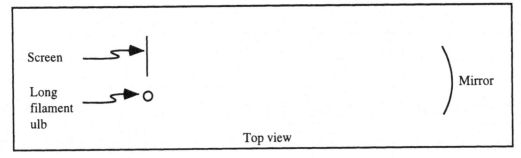

A. Predict what would happen to the image if the left half of the mirror were covered by a mask. Explain your reasoning.

B. Predict what would happen to the image if a mask with a small hole in the center (smaller than the object) were placed in front of the mirror.

How would your answer change if:
- the hole is made larger than the object
- the hole is moved so it is not at the center of the mirror

C. Perform each of the above experiments and check your predictions. If your predictions were incorrect, draw new ray diagrams that are consistent with what you observe.

McDermott & P.E.G., U.Wash./*Physics by Inquiry*

Section 12. Applications of geometrical optics

The phenomena of reflection and refraction are the basis of a number of optical

instruments. In this section, we investigate several common optical instruments.

Experiment 12.1

Imagine that you wanted to spy on your neighbors who had erected a tall
fence at your property line. How could you use two mirrors and a
cardboard tube to create a *periscope,* a device that you could use to see
over the fence?

Design and construct such a device.

Experiment 12.2

Obtain a plane mirror, a convex mirror, and a concave mirror. If needed,
partially cover the mirrors so all three mirrored surfaces are the same size.

Hold each mirror one at a time in one hand at arm's length from you.
Look at the images of objects behind you.

A. Which mirrors produce upright images?

Which mirrors produce images that appear larger than, smaller than, or
the same size as the objects? Explain.

B. Is the extent of what you can see behind you (i.e., your field of view)
different for the three mirrors? Explain.

C. Which type of mirror might be best suited for use as a *rear-view
mirror?* Explain.

What sort of warning would you give drivers who use the mirror?

Experiment 12.3

A *magnifying glass* is a convex lens used to make an object appear larger
than it is.

A. Use a convex lens as a magnifying glass.

Is the image erect or inverted? Is it real or virtual?

Does the lens make the object appear closer to you or farther from
you? Use parallax to determine where the image is located.

B. Sketch a ray diagram to illustrate a lens used as a magnifying glass. The diagram need not be drawn to scale but should show the correct object and image locations with respect to the lens and its focal points.

Does a magnifying glass simply make objects appear closer? If not, what does it do?

✔ Discuss your answers with a staff member.

Exercise 12.4

In a *camera,* light from an object reaches a piece of film and exposes it.

A. Describe how to use a convex lens to make a simple camera. Does it matter whether the image is real or virtual?

Make a sketch showing the location of the convex lens, its focal points, the object and image locations, and the location of the film.

B. Imagine that you had just taken an in-focus photograph of a distant object. Would you need to adjust your camera to take a photograph of something much closer? If so, how? Explain.

Explain why in some photographs, certain objects are in focus while other objects are not.

C. Cameras use a shutter to control the length of time the film is exposed (e.g., 1/60th of a second). The more light that reaches a point on the film, the brighter the corresponding point on the photograph.

How would different exposure times affect the photograph?

When taking a photograph in a dim room, what would be an advantage of using a long exposure time?

When taking a photograph of a quickly moving object, why might you want to use a short exposure time?

D. In *Light and Color,* Volume I, you studied pinhole cameras. If you use a pinhole camera to take a photograph, where should you place the film?

If you have just used a pinhole camera to take a photograph of a distant object, do you need to adjust the camera to take a photograph of something much closer? If so, how would you adjust the camera?

In part B, it was noted that in some photographs, certain objects are in focus while other objects are not. Would this be true of a photograph taken by a pinhole camera?

Describe two ways that increasing the size of the hole in a pinhole camera would affect a photograph.

Why might a pinhole camera be more appropriate for taking pictures of stationary objects rather than moving objects?

E. Cameras typically include a *diaphragm* (an aperture or opening of adjustable size). This opening is near the lens between the lens and the film.

What would be the effect on the image of making the aperture larger or smaller?

Experiment 12.5

An eye can be modeled as a convex lens near a detector called the *retina*.

A. Would you expect a retina to detect real and virtual images equally well? If not, would you expect the lens in the eye to form real images or virtual images? Where do you think the images would be located?

B. In order to focus on objects that are at various distances, the focal length of the lens in the eye can change, while the distance between the lens and the retina is relatively unchanged.

How does this situation differ from that for a camera?

If you were to focus on something farther away than the words on this page, would the focal length of your lens increase or decrease? Draw ray diagrams to support your answer.

Estimate the focal length of the lens in your eye when you are viewing a distant object. (Assume the retina is at the back of the eyeball, and assume a reasonable diameter for the eyeball.)

C. Not everyone can focus on distant objects. In a *farsighted* eye, light from distant objects converges too far from the lens; in a *nearsighted* eye, too close.

Obtain a paper screen to represent a retina, a convex lens to represent a lens, and a lighted bulb to represent a distant object. Use these materials to illustrate a normal eye viewing a distant object.

Change your arrangement to illustrate: (1) a farsighted eye viewing a distant object and (2) a nearsighted eye viewing a distant object.

D. Do you think you would be able to correct farsightedness using glasses made from a convex lens? Test your answer by using the equipment from part C and a second convex lens.

Do you think you would be able to correct farsightedness using glasses made from a concave lens? Test your answer by using the equipment from part C and a concave lens.

E. Repeat part D, attempting to correct nearsightedness.

F. Describe an experiment you could perform to determine whether a pair of eyeglasses corrects for farsightedness or nearsightedness. If you or your partner wears eyeglasses, test your experiment.

Experiment 12.6

In this experiment, we construct a simple *telescope,* a device to magnify an object that is far from you.

A. Use a convex lens to form an image of a distant object, such as a window or lighted bulb.

Is the image erect or inverted? Is the image real or virtual?

B. Place a thin paper screen behind the lens so that an image of the object appears on the screen. Obtain a second convex lens with a smaller focal length than the first. Use the second lens as a magnifying glass to examine the back of the screen.

Remove the screen while you are looking at the image on the screen. Describe your observations.

Is the image that you see real or virtual? Is it erect or inverted?

In simple terms, what is the purpose of the second lens?

In simple terms, what is the purpose of the first lens?

C. Draw a ray diagram that illustrates the use of the two lenses as a telescope. Your diagram must include both lenses, however, you may find it easier to consider the lenses and images one at a time.

What would be one disadvantage of replacing the first lens by a lens with the same focal length, but with a smaller diameter?

Experiment 12.7

In this experiment, we construct a simple *microscope,* a device to magnify an object that is close to you.

A. How would you use two convex lenses to make a *microscope?*

Use two lenses of short focal length to construct such a device.

In simple terms, what is the purpose of each lens?

B. Draw a ray diagram that illustrates the use of the two lenses as a microscope. Your diagram must include both lenses, however, you may find it easier to consider the lenses and images one at a time.

✔ Discuss the preceding two experiments with a staff member.

     McDermott & P.E.G., U.Wash./ *Physics by Inquiry*

Experiment 12.8

Obtain a plane mirror, a convex mirror, and a concave mirror. If needed, partially cover the mirrors so all three mirrored surfaces are the same size.

One at a time, place each mirror close to a small lighted bulb. Move the bulb toward and away from each mirror and record your observations.

Which mirror might be best suited for use in a *headlight?* Explain the important features of your design with words and using a sketch.

Experiment 12.9

Obtain a prism, a beaker of water, and a transparent sphere. Shine a very narrow beam of light through each of these objects onto a sheet of white paper. Record your observations.

A. For each of the materials you are using, does red light bend more than, less than, or the same amount as blue light? Explain how you can tell from your observations.

B. How are your observations of refraction of white light by the prism, water, and transparent sphere similar?

C. Construct a model for the appearance of a rainbow. (*Hints:* Have you seen a *rainbow* in the sky on (1) a sunny day, (2) a cloudy day, and (3) a rainy day? When you see a rainbow, what is the light source? What are the refracting objects?)

Exercise 12.10

In the preceding experiment you considered how the index of refraction of a material depends on the color of the incident light. In this exercise, we consider how this behavior affects the ability of a lens to focus light.

A. Consider an object a distance s from a convex lens. Imagine that the object is first illuminated by red light, and then by blue light.

Will the image location change? If so, in which case will the image be closer to the lens? Explain your reasoning.

B. Repeat part A, considering a concave mirror instead of a lens.

C. Suppose that the object is now illuminated with white light and again placed in front of the lens.

Would the image be located at a unique position or would it appear fuzzy? Explain your reasoning.

This effect is called *chromatic aberration.*

Experiment 12.11

A. Obtain an *optical fiber* (i.e., a long, thin, transparent strand of plastic or glass. Observe both ends of the fiber).

   Place one end of the fiber close to a source of light. Keep the fiber straight. Describe the appearance of the other end of the fiber.

   Keep one end of the fiber close to the source of light. Bend the fiber and describe the appearance of the other end.

B. To account for your observation that the direction of light appears to change, consider the enlargement of a section of the fiber below.

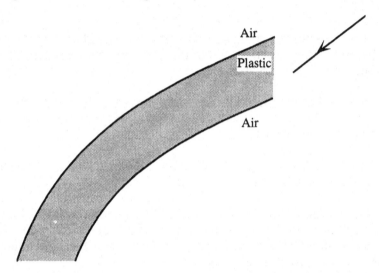

   Continue the light ray that enters the fiber from the right until it reaches the plastic-air interface. Is the light ray reflected, transmitted, or both when it reaches this interface?

   Sketch the light ray after it has reached the plastic-air interface and continue the path until the light leaves this section of the fiber.

   Use the ideas from your drawing to account for the behavior of light in an optical fiber. Explain your reasoning.

C. This part of the experiment must be performed in a darkened room. Obtain a large funnel, a bright flashlight, a sheet of white paper, and either a sink or a bucket, and some water. Place your hand over the bottom of the funnel and fill the funnel with water. Hold the funnel so that the water will flow out at an angle rather than straight down. Aim the flashlight through the funnel as the water flows out onto a piece of white paper in the sink.

   Note where the light beam hits the paper and compare it to where the light beam hits when the funnel is empty.

   Use the ideas in part B to explain why the light hits the paper in different locations. Use a drawing in your explanation.

Supplementary problems for *Light and Optics*

Problem 1.1

A barrier is placed in front of, but not touching, a plane mirror. The barrier is perpendicular to the mirror. Suppose you have a flashlight that projects a narrow beam of light. A target is placed successively at positions A, B, and C on the left side of the barrier.

A. For target locations A and B, indicate a location on the right side of the barrier where the flashlight could be placed in order to hit the target with light. Explain how you determined these locations.

B. For target location C, determine the entire *region* on the right side of the barrier in which a flashlight could be placed in order to hit the target with light. Explain how you determined the region.

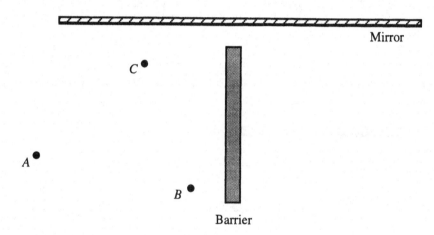

Problem 1.2

After class one day, you find a sheet of paper on the floor with the following drawing on it. Evidently, a student was doing an experiment that used mirrors and a light source, but the student apparently forgot to indicate the locations of the mirrors on the paper. On the diagram, indicate the locations and orientations of the mirrors and explain how you determined the arrangement of the mirrors used by the student.

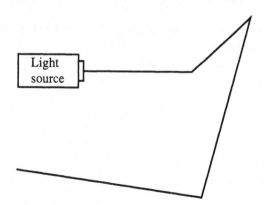

McDermott & P.E.G., U.Wash./ *Physics by Inquiry*

Problem 1.3

Two students are sitting near a wall from which hangs a covered mirror. In front of the mirror is a wooden dowel on a stand.

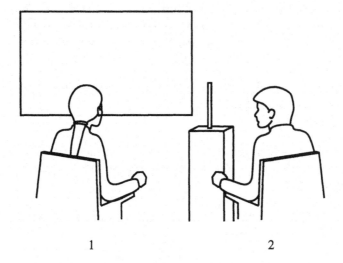

The diagram below is a top view representation of the situation pictured above.

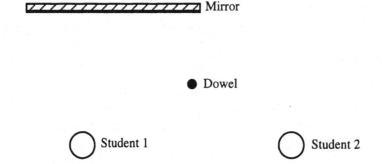

A. In what order (left to right) will the following appear to student 1: his own image in the mirror, student 2's image, the dowel's image, and the dowel itself? If Student 1 can not see his image, student 2's image or the dowel's image, state that explicitly.

B. In what order (left to right) will the following appear to student 2: her own image in the mirror, student 1's image, the dowel's image, and the dowel itself? If Student 2 can not see her image, student 1's image or the dowel's image, state that explicitly.

Problem 1.4

An observer and a cube are located in front of a large mirror as shown in the top view diagram below. The cube has a narrow stripe down the middle of two sides.

From the location shown, could the observer see:

(1) the image in the mirror of stripe 1? Explain.

(2) the image in the mirror of stripe 2? Explain.

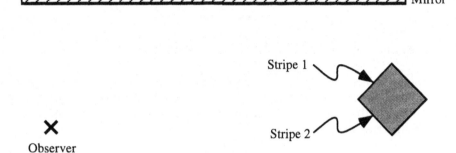

Problem 1.5

The ballroom shown below has two large mirrors mounted on the ceiling.

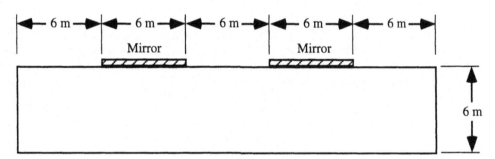

A. In what region of the floor could a small mouse be located so that when it looks at the mirrors it could see another mouse on the floor at either end of the room? Explain your reasoning.

B. Is there anywhere on the floor that a mouse could be located and see the entire ballroom floor in the mirrors? Again, explain your reasoning.

C. Would your answer to part B differ if the ceiling were higher, but the length of the room didn't change? Explain.

McDermott & P.E.G., U.Wash./ *Physics by Inquiry*

Problem 1.6

The line *AB* is parallel to the plane of a mirror. The mirror is 6 meters away from the line and is *L* meters long.

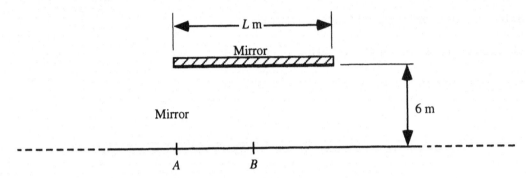

Explain your reasoning in each part of the question below.

A. A man stands with his eyes at point *A*, directly under one end of the mirror,. How much of the line can he see in the mirror? If the man stands with his eyes at point *B*, under the center of the mirror, how much of the line can he see? In each case, express your answer in terms of *X*.

B. Does the man's location along the line affect how much of the line he is able to see?

C. How, if at all, would your answers to parts A and B differ if the mirror were moved so it is twice as far from the line?

Problem 2.1

When you are sitting in a chair, it appears to you that object A is above object B. When you stand up, object A appears to be below object B. Which of the two objects is farther away from you? Draw a diagram showing the locations of the two objects and your eye, and explain these observations.

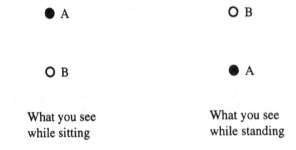

McDermott & P.E.G., U.Wash./ *Physics by Inquiry*

Problem 2.2

In each case below, you are given two lines of sight to the image of an object in a plane mirror. In addition, you are told where the object is.

For each case determine whether this situation is possible. If so, draw the location and orientation of the mirror.

Explain how you reached your conclusions.

(1)

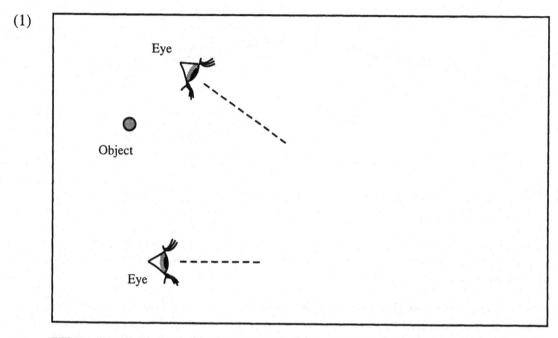

(2)

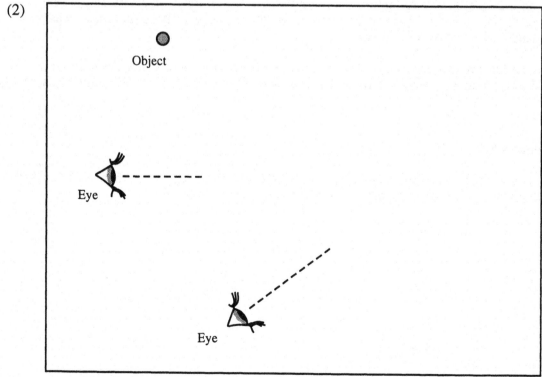

    McDermott & P.E.G., U.Wash./ *Physics by Inquiry*

Problem 3.1

Two mirrors are placed at an angle of 160° as shown.

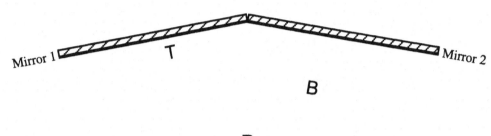

A. Using parallax, locate and draw the images of the letters *T, R,* and *B.* (Use real mirrors in this part.)

B. Give a detailed explanation of what you see. Identify each image as either a perverted or a non-perverted image. For each image, give the sequence of mirrors on which reflections occur.

C. Show some representative light rays that indicate how two of the images are formed. Pick a pair of rays that form an image after one reflection and another pair that form an image after two reflections (if there are any). You may select any single point on each object as the origin of the rays. Draw the rays carefully so that they clearly satisfy the law of reflection.

D. Is there a region in front of the mirrors that is "special" in terms of the number of images that are possible? If so, describe where that region is and tell why it is special.

Problem 3.2

Show that for two mirrors at an angle of 90° to each other, a light ray that reflects off of both mirrors follows a path that is parallel to the incident ray.

Problem 4.1

The following are top view diagrams of a beaker of water. Some of the diagrams represent qualitatively correct paths for light through a beaker of water and some do not. Which diagrams have "flaws"? Briefly explain the flaw in each case.

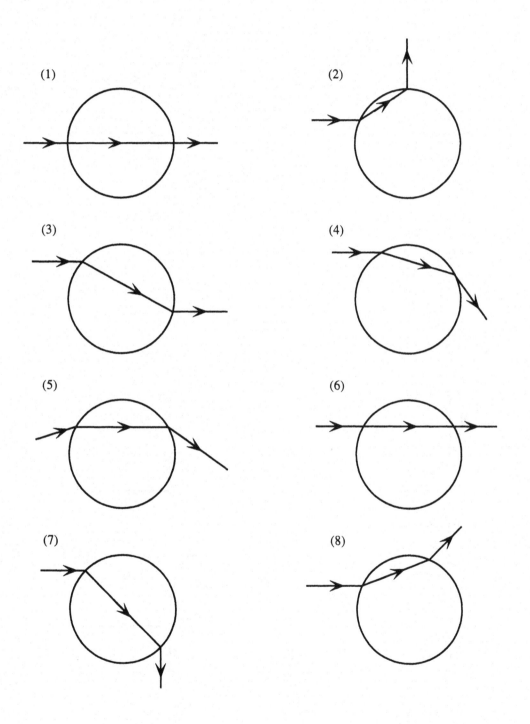

(1)

(2)

(3)

(4)

(5)

(6)

(7)

(8)

McDermott & P.E.G., U.Wash./*Physics by Inquiry*

Problem 4.2

The following are top view diagrams of a container of water that has flat sides. Some of the diagrams represent qualitatively correct paths for light through the container of water and some do not. Which diagrams have "flaws"? Briefly explain the flaw in each case.

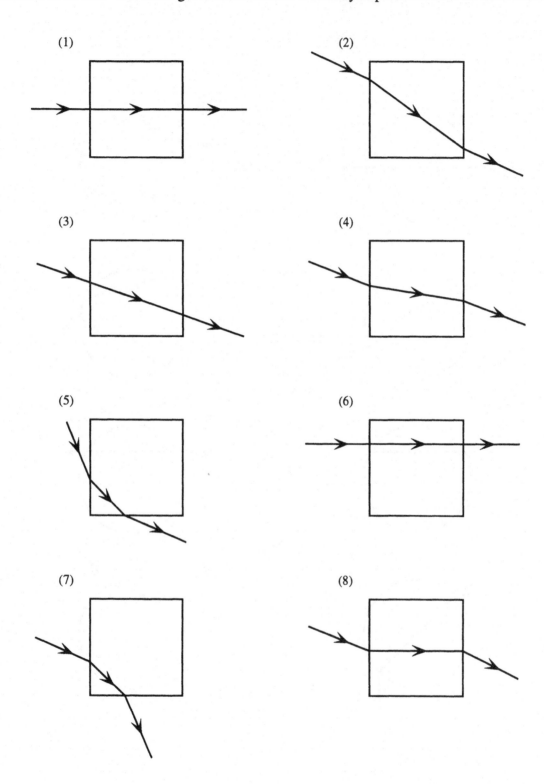

Problem 4.3

After class one day, you find a sheet of paper with the following drawings on it. On the table nearby, there is a light box and several containers of water. There are both round-bottomed beakers and square-bottomed plastic containers, but *no* semi-cylindrical containers. Evidently, a student was doing an experiment that used the containers of water and the light box, but the student apparently forgot to indicate the locations of the containers on the paper.

A. Draw an outline of each of the containers on the diagram for the student. (Your final diagrams should be *qualitatively* correct. You may find it helpful to read part B first.)

B. For each of the container locations that you indicated, can you be certain whether a student was using a container with curved sides or flat sides? Are there any that must be curved? Are there any that must be flat? Are there any that could be either? *If a container could be either type, show both possibilities on the diagram.*

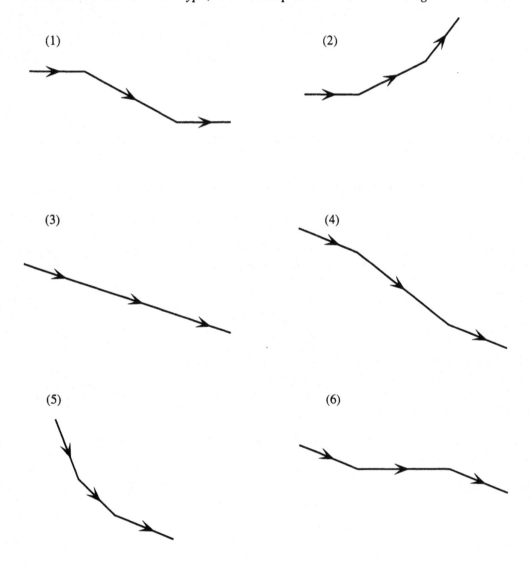

(1)

(2)

(3)

(4)

(5)

(6)

McDermott & P.E.G., U.Wash./ *Physics by Inquiry*

Problem 4.4

A piece of cardboard is placed between a very small light bulb and a screen. A block of glass is placed in front of the screen as shown in the side view diagram at right.

If the glass were *removed,* would the size of the shadow on the screen increase, decrease, or stay the same? Explain your reasoning, and support your answer with a clear, qualitatively correct ray diagram.

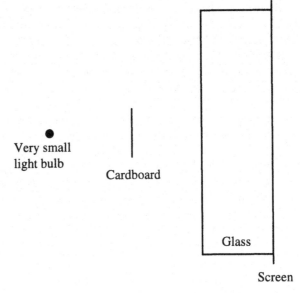

Problem 4.5

A. Is it possible to look into a mirror and see your friend's eyes, yet your friend cannot see your eyes? Explain.

Does it matter if the light strikes several mirrors before reaching your eyes?

Does it matter if the light passes through one or more beakers filled with water?

B. Imagine that you are underwater, and your friend is out of the water. If you can see your friend's eyes, can your friend see your eyes? Explain.

Problem 5.1

A beam of light enters a plastic cube of side 3 cm from the center of the left side as shown in the diagram at right. The angle of incidence is 75°. The index of refraction of the plastic with respect to air is 1.5.

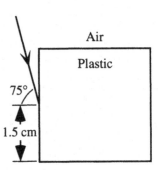

A. From which side of the cube will the ray emerge? Draw a diagram to support your answer.

B. At what angle will the ray emerge?

McDermott & P.E.G., U.Wash./ *Physics by Inquiry*

Problem 5.2

One day, a student determines that light from the
sun strikes the ground at an angle of 50° from the
vertical. The student is 1.6 m tall.

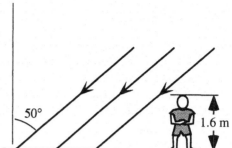

A. What is the length of the student's shadow?
Explain.

B. The student then goes completely underwater
on the bottom of the deep end of a large swimming pool. The surface of the pool is
still and the bottom of the pool is level. What is the length of the student's shadow?
Show your work.

Problem 5.3

Secure a water-resistant watch to your wrist and place in a sink of water so that the face
of the watch is toward your eyes. Move your wrist so that the watch slowly rotates away
from your face.

Is there an angle past which you can continue to see the face of the watch but you can
no longer see the hands (or the digital display) of the watch?

How can you account for your observation? Draw a diagram to show the path of light
rays from the hands (or the digital display) of the watch through the flat transparent
face, through the water, to your eyes. How does your diagram change as the watch is
rotated?

Is the index of refraction of the transparent face plate of the watch greater than, less
than, or equal to that of water? Explain.

Problem 6.1

A hungry fish in a tank of water notices a small
fly that is located directly above the fish's eye as
shown at right. Draw a clear, qualitatively
correct ray diagram that shows the apparent
location of the fly *as seen by the fish* (i.e., the
image location of the fly).

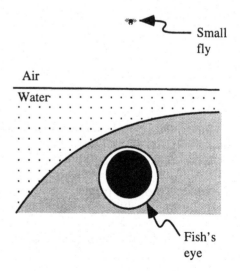

McDermott & P.E.G., U.Wash./ *Physics by Inquiry*

Problem 6.2

On an extended backpacking trip, you find that a bridge across a river has been washed out during a flash flood. The only way back to civilization is to wade across the river. How well can you judge depth of the river visually? Will you get wetter than you expect, not as wet as you might expect, or can you judge the depth of the river perfectly? Explain your reasoning and draw a ray diagram that supports your answer.

Problem 6.3

An observer is looking at a pin at the bottom of a container of water as shown in the side view diagram at right. Draw an *accurate* ray diagram to determine where the pin appears to be located to the observer (i.e., the image location). Use a protractor and a straightedge in drawing your diagram. Use the value that you determined for the index of refraction of water relative to air.

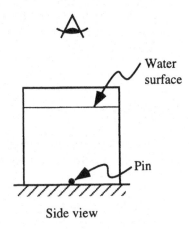

Water surface

Pin

Side view

Problem 7.1

You have two convex lenses: one with focal length f and one with focal length $10f$. Which lens would you say is "stronger"? Explain your reasoning.

Problem 7.2

What are two simple ways you could use to estimate the focal length of a lens? Explain your reasoning.

Problem 7.3

In Experiment 7.4, you observed that when a screen was placed at the image location, an image was formed on the screen. However, when the screen was moved toward or away from the lens, the image quickly disappeared.

A. Discuss how an image was formed on the screen when it was held at the image location. Include a ray diagram in your discussion.

B. Explain, both in words and with a ray diagram, the change in the image as the screen was moved.

Problem 8.1

A pencil is placed a distance $2f$ to the left of a convex lens, where f is the focal length of the lens. A student has found that the image is located $2f$ to the right of the lens, as shown in the scale diagram below.

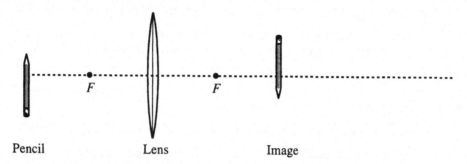

Pencil   Lens   Image

A. Determine the region where an observer's eye must be located in order to see the *entire* image of the pencil. Clearly label this region on your diagram.

B. For an observer who can see the entire image, does the image *appear* longer than, shorter than, or the same length as the pencil would without the lens? Explain.

Problem 8.2

A. A small object is placed in front of a convex lens as shown below. Draw an accurate ray diagram using all principal rays to determine the location of the image. Explain how you determined the image location. Label each of the rays, and in words, explain how you decided to draw each ray in the way that you did.

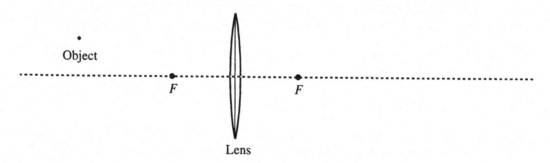

Lens

B. Repeat part A for the object closer to the lens as shown below.

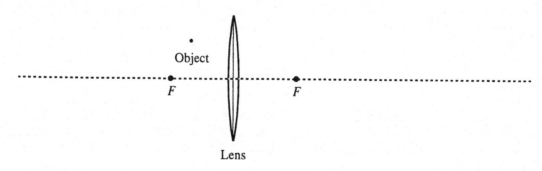

Lens

Problem 9.1

An object is placed in front of a convex lens. The focal length of the lens is 20 cm.

A. Suppose the object is 60 cm from the lens.

> On a sheet of graph paper, draw a ray diagram to find the location of the image.

> Use the thin lens equation to calculate the location of the image. Is your answer consistent with your ray diagram? If not, resolve any discrepancies

> Is the image real or virtual? Explain.

B. Suppose the object is 10 cm from the lens.

> On a sheet of graph paper, draw a ray diagram to find the location of the image.

> Use the lens equation to calculate the location of the image. Is your answer consistent with your ray diagram? If not, resolve any discrepancies

> Is the image real or virtual? Explain.

Problem 9.2

A light bulb is 100 cm from a screen. Suppose you have a convex lens with a focal length of 21 cm. At what location(s) between the bulb and the screen could you place the lens such that the image of the bulb would be on the screen? Show your work.

Problem 9.3

A student walks into a room and sees a lighted bulb 45 cm in front of a convex lens. He notices that there is a screen 90 cm from the lens with an image of the bulb on the screen.

A. Are the bulb and the screen on the same side or of the lens or on opposite sides of the lens? Explain.

B. Determine the value of the focal length of the lens.

C. Suppose the screen were moved 20 cm farther from the lens. Would you be able to position the bulb so that you could see a clear image of the bulb on the screen? If so, where? If not, why not? Explain.

D. Suppose that instead the screen were moved 70 cm closer to the lens than its original position. Would you be able to position the bulb so that you could see a clear image of the bulb on the screen? If so, where? If not, why not? Explain.

Problem 9.4

In Exercise 9.6, you determined an expression for the magnification of a real inverted image formed by a convex lens. Does this expression hold for virtual images as well? If not, what is the expression in this case? Explain your reasoning, and support your answer with a ray diagram.

Problem 9.5

A convex lens is used to create a real image of a pencil. The locations of the pencil and its image are shown below. The lens, however, is not shown.

Pencil   Image

A. Where is the lens located? Explain how you determined your answer.

B. Where are the focal points of the lens located? Explain how you determined your answer.

Problem 9.6

A convex lens is used to create a real image of a small object. The diagram below shows the locations of the lens, its focal points, and the image. Determine the location of the object. Explain how you determined your answer. Treat the object as a point source of light.

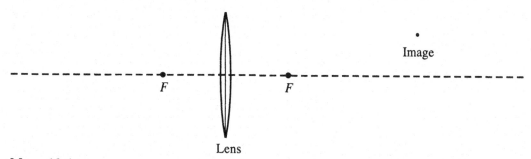

F   *F*

Image

Lens

Problem 10.1

A. A small object is placed in front of a concave lens as shown below. Draw an accurate ray diagram to determine the location of the image. Label each of the rays, and in words, explain how you decided to draw each ray in the way that you did. Treat the object as a point source of light.

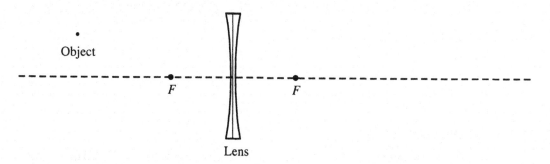

Object

F   *F*

Lens

B. Repeat part A assuming the lens is replaced by a convex lens with the same magnitude focal length.

McDermott & P.E.G., U.Wash./ *Physics by Inquiry*

Problem 10.2

A. A small object is placed in front of a concave lens as shown below. Draw an accurate ray diagram to determine the location of the image. Label each of the rays, and in words, explain how you decided to draw each ray in the way that you did. Treat the object as a point source of light.

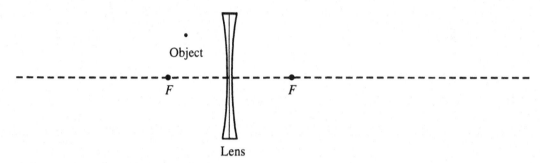

B. Repeat part A assuming the lens is replaced by a convex lens with the same magnitude focal length.

Problem 10.3

A. The diagram below shows a small object and its image, which is virtual. The principal axis of the lens is shown, however the lens is not. Treat the object as a point source of light.

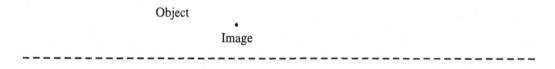

Is this situation possible?

    If so: Is the lens concave, convex, or is either type possible? Determine the locations of the lens and its focal points. Explain how you determined your answers.

    If not: Explain why not.

B. Suppose the object and image locations from part A were reversed.

Is this situation possible?

    If so: Is the lens concave, convex, or is either type possible? Determine the locations of the lens and its focal points. Explain how you determined your answers.

    If not: Explain why not.

McDermott & P.E.G., U.Wash./*Physics by Inquiry*

Problem 10.4

In Exercise 9.6, you determined an expression for the magnification of a real inverted image formed by a convex lens in terms of the object and image distances. Does this expression hold for images formed by concave lenses as well? If not, what is the expression in this case? Explain your reasoning, and support your answer with a ray diagram.

Problem 11.1

Consider the ray drawn incident on the mirror in the diagram below. Determine the continuation of the ray. Explain your reasoning.

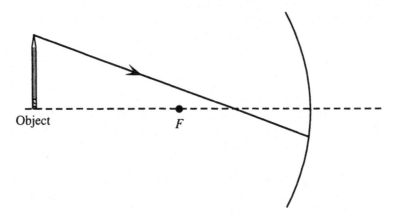

Problem 11.2

A mirror placed to the right of a pencil is used to create a virtual image of it. The locations of the pencil and its image are shown below. The mirror, however, is not shown.

A. Is the mirror plane, convex, or concave? Explain how you determined your answer.

B. Where is the mirror located? Explain how you determined your answer.

Problem 11.3

In Exercise 9.6, you determined an expression for the magnification of a real inverted image formed by a convex lens in terms of the object and image distances. Does this expression hold for images formed by curved mirrors as well? If not, what is the expression in this case? Explain your reasoning, and support your answer with a ray diagram.

McDermott & P.E.G., U.Wash./ *Physics by Inquiry*

Problem 11.4

A flower and an empty vase are placed in front of a concave mirror. The flower is not shown in the diagram at right.

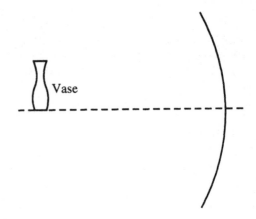

A. Where should the flower and the vase be placed so that the image of the flower is same size as the actual flower *and* the image of the flower is located in the actual vase as seen by an observer on the principal axis of the mirror? Draw a ray diagram to support your answer.

B. Determine the image and object distances in terms of the focal length of the mirror.

Problem 12.1

You have observed that if the angle of incidence of a beam of light striking a mirror takes on a very special value (i.e., zero), the beam is reflected so that it reverses its original direction. However this is only true for this one particular direction of the incident light.

It is possible to use several plane mirrors to design a *reflector,* a device that reverses the direction of any beam of light that strikes it—independent of the relative orientation of beam and reflector. Make a sketch of the reflector and draw ray diagrams to illustrate how it works. (*Hint:* This instrument is sometimes called a corner reflector.)

Problem 12.2

Consider a parallel beam of light of a given color striking a spherical water droplet as illustrated in the cross-sectional top view diagram at right. The index of refraction of water with respect to air is 1.33 for that color.

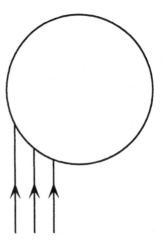

A. Continue each of the rays shown through and out of the droplet. Use a ruler and a protractor to draw an accurate diagram.

B. Suppose a student were looking at the droplet and the light source was behind her. Would she be able to see any of the emerging rays? Explain.

C. What would you expect the relative orientation of an observer, the sun, and rain drops to be during the appearance of a rainbow? Check your answer the next time you see a rainbow.

McDermott & P.E.G., U.Wash./*Physics by Inquiry*

PHYSICS BY INQUIRY

Kinematics

In *Kinematics,* we study how motion in one dimension can be described in terms of the concepts of position, displacement, velocity, and acceleration. Graphical and algebraic representations are introduced and we investigate how these can be used to represent and predict the motion of a real object.

Part A: Motion with constant speed

We begin our investigation of *Kinematics* by studying the motion of objects that move in a straight line without speeding up or slowing down. This kind of motion is called *uniform motion* or *motion with constant speed.*

Section 1. Uniform motion

Experiment 1.1

Obtain several balls and tracks from the staff.

A. Using this equipment, try to produce the most uniform, steady motion you can.

    Describe what you did to make the motion as uniform as possible.

    How can you check whether the motion was uniform? Explain why you believe your method is a good test for uniform motion.

    Give quantitative evidence that the motion is uniform.

B. Write an operational definition of uniform motion. (You may need to refer to the discussion of operational definitions in *Properties of Matter* in Volume I.)

    Note: Such a definition should include a test that can be used to decide whether a motion is uniform. In principle, the test should be one that could be applied to any moving object. It should not involve personal judgments about how uniform the motion appears. Instead, the test should require making measurements of the motion. Then there should be a procedure for using these measurements to decide whether or not the motion is uniform.

✔ Discuss this experiment with a staff member.

Experiment 1.2

Gather together a group of four or more people for this experiment.

The purpose of this experiment is to keep track of where a moving object is at various times. This can be done by noting the time when the object passes each of several checkpoints.

Set up a track at least two meters long so that a ball rolls along the track with nearly uniform motion.

A. Each person should be assigned a checkpoint along the path of the moving ball. What is a good way of describing the location of a checkpoint?

   Each person should get a stopwatch or stop clock and start it when the ball passes the starting point. Each person's clock should be stopped when the ball passes his or her checkpoint.

   What will the clock readings tell you?

   Practice a few times until you can repeat the experiment the same way each time. When you are able to obtain the same clock readings repeatedly, take your data.

   Is the motion uniform? How can you tell? If the motion is not uniform, change your apparatus to make the motion more uniform.

B. When you have data for a particular run of the experiment indicating that the motion was uniform, plot your data on a graph. (Use only the data from your best run.) Make the following plot: distance from the start to the checkpoint versus clock reading at the checkpoint. *Use graph paper!*

   Note: When experimental data points do not lie exactly in a neat pattern, the usual practice is to draw a smooth curve or line through the points. *Do not draw a connect-the-dots, zigzag line.* Such a line implies that you think the measured data points are exact, with no uncertainty. A zigzag graph indicates that the quantities are related in a very irregular way; a smooth graph indicates a more regular relation.

C. The points should nearly form a straight line. Draw the best straight line that you can to represent your data. What does the straightness of the graph tell you about the motion?

   Compute and interpret the slope of the line.

D. Plot the same data on another graph, this time with the axes reversed. Again compute and interpret the slope.

✔ Check your graphs and interpretations with a staff member.

Experiment 1.3

Produce a motion in which an object moves uniformly through 2.0 meters in 4.0 ± 0.2 seconds. Repeat Experiment 1.2 using this motion. Plot your results on the same graphs you made for Experiment 1.2.

Compare the graphs for Experiments 1.2 and 1.3. How can you tell which motion was faster by glancing at the graphs?

Experiment 1.4

Use the same setup as in the previous two experiments, but this time use only two clocks. Place one clock facing you at the 120 cm mark. Place the other clock at the 70 cm mark, but turn its face so that you cannot see it. Run the experiment using a fairly slow speed you have not used before.

Read the clock at the 120 cm mark and then try to predict the reading of the hidden clock at the 70 cm mark. Explain your reasoning. Then see whether your prediction was correct.

Experiment 1.5

Perform an experiment like Experiment 1.2 but with an object that is speeding up.

A. Plot a graph of your results. Follow the accepted convention of recording distance from the start on the vertical axis and clock readings on the horizontal axis.

   Is it proper to try to draw a straight line through your points? Explain.

   How can you tell that the object is speeding up by looking at the graph?

B. Try to make a prediction as in Experiment 1.4. Can you use the same reasoning here? Explain.

✔ Discuss this experiment and Experiment 1.4 with a staff member.

McDermott & P.E.G., U.Wash./ *Physics by Inquiry*

Section 2. Quantitative descriptions of positions and times

Whenever an object moves along a straight line, it is possible to imagine a meterstick lying alongside the path of the object. We can imagine this meterstick as extending forever in both directions. The meterstick represents a *coordinate system*.

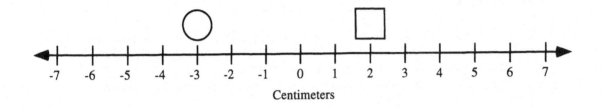

We can describe where an object is by stating the number that is closest to it along the meterstick. This number is called the *position* of the object. We use the symbol x to represent position. In the figure above, the square is next to the +2 cm mark, so it has position +2 cm. Using the symbol x, we would say that $x = 2$ cm for the square. For the circle, $x = -3$ cm.

Displacement

If an object moves from an initial position x_i to a final position x_f the number $x_f - x_i$ is called the *displacement* or the *change of position* of the object. We will use the symbol Δx for $x_f - x_i$: $\Delta x = x_f - x_i$, where Δx is read "delta x." In this module, we always use the Greek letter *delta* (Δ) to mean final value minus initial value. Thus the symbol Δ represents a change in a quantity.

A displacement is often represented by an arrow drawn from the initial position to the final position. The length of the arrow represents how far the final position is located from the initial position, and the arrowhead shows the direction of motion. An arrow representing displacement is shown in the following figure.

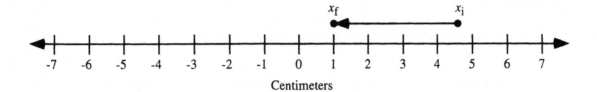

Exercise 2.1

Calculate Δx for the preceding figure. What does the sign of Δx tell you about the motion?

Exercise 2.2

Calculate the displacement for each of the following motions. In each case, x_i is the first position and x_f is the later position. Draw an arrow to represent each displacement as in the preceding figure.

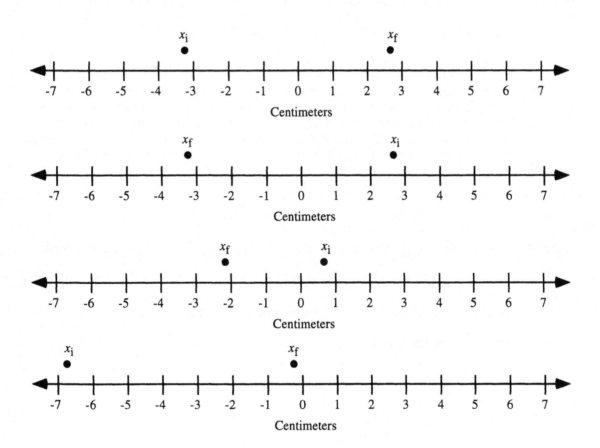

Exercise 2.3

Which quantity tells where an object is located, x or Δx?

If an object keeps moving in the same direction, $|\Delta x|$ is the distance the object moves. If the object turns around, however, the distance it travels is greater than $|\Delta x|$. (The symbol $|\Delta x|$ means the *absolute value* of Δx. Consult a staff member or an algebra text if you are not familiar with the term *absolute value*.)

$|\Delta x|$ is the distance traveled. $|\Delta x|$ is not the distance traveled.

Starting positions

Motion may begin at any position number. Often $x = 0$ is defined to be the place where the motion begins, but this is not necessary. Consider the following example:

A farmer lives between two large cities, A and B. The highway that passes the farm is marked with mileposts. The numbering system starts at city A and continues on to city B. The farm is located at milepost 198. Consider a trip the farmer took into the nearby small town at milepost 206. Her initial position according to the highway mileposts was not 0, since she did not start the trip at city A. Her initial position was $x_i = 198$, and her final position was $x_f = 206$. The farmer might live her entire life without ever going to city A. In that case, her position would never be 0.

We will frequently encounter situations like the case above, where the motion does not start at $x = 0$. The choice of location for $x = 0$ is arbitrary; however, once the choice has been made for a particular problem, it should not be changed. The location where the position is zero is often called the *origin* of the coordinate system. The following exercise illustrates how the choice of the origin affects position and displacement measurements.

Exercise 2.4

Two students conduct an experiment in which an object is moved from
position x_i to position x_f. Student 1 and student 2 each place a meterstick
next to the line of motion, as shown below.

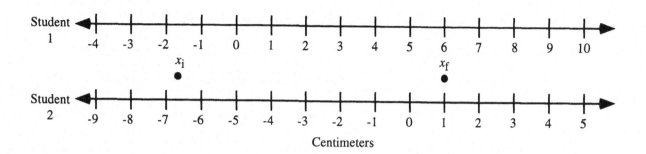

What value would student 1 measure for x_i?

What value would student 1 measure for x_f?

What value would student 2 measure for x_i?

What value would student 2 measure for x_f?

What value would student 1 give for Δx?

What value would student 2 give for Δx?

Answer the following questions on the basis of your results above.
Explain your reasoning.

Does the measurement of position depend on the choice of coordinate
system?

Does the measurement of displacement depend on the choice of
coordinate system?

Time

The word *time* has several meanings in ordinary language. For example: What time did

he arrive? How much time does that job take? We had a fine time. Time out. Our time

has come. In physics, the word *time* must be used carefully because it is used to name

several quantities. The exact meaning of the word in physics is often told in the

accompanying words, for example, *at that time, for a long time.* We will be concerned

with two uses of the word corresponding to two different quantities: one called *the time,* meaning a clock reading, and the other called *duration, time interval,* or *amount of time.* These quantities are discussed below.

In describing position, we imagine a meterstick that continues forever in both directions. When considering time, we imagine a clock that always has been running and will continue to run forever. Instead of an ordinary clock, we imagine a clock that just keeps count of seconds. It does not start over after 12 hours as our usual clocks do. The clock we imagine keeps counting up to millions and billions of seconds. We use the symbol t to represent the clock's reading.

The number on the clock is called the *time.* We describe when something happens by giving the time shown on the clock (the clock reading) as the event occurs. A typical event, for example, would be the passing of the 10 cm mark by a moving object. If the clock read 25 seconds as this happened, we would say that $x = 10$ cm at $t = 25$ seconds.

Just as $x = 0$ is not necessarily a special place, $t = 0$ is not necessarily a special instant. In particular, $t = 0$ does not have to be the instant motion begins. For example, suppose two cars are making the same trip but one of them has a five-minute head start. We might set our clock to read $t = 0$ when the first car starts, but then the second car's motion would begin at $t = 5$ minutes.

Instants before $t = 0$ have negative time numbers, such as $t = -5$ seconds. This is similar to what we do for years before the year one, such as 1026 B.C. Negative times are perfectly ordinary; they are just earlier than positive times.

Exercise 2.5

Suppose a ball was dropped at $t = 2140.0$ seconds and hit the floor at $t = 2141.2$ seconds. How much time did it take the ball to fall?

The problem above illustrates the method of determining duration or amount of time between two instants. We use the symbol Δt to mean $t_f - t_i$, the amount of time between t_f and t_i. Again, Δt always means "final time minus initial time," and is read "delta t."

Exercise 2.6

Two observers timed the motion of a car from one place to another. The first observer's clock read 262 seconds at the start and 375 seconds at the end of the car's motion. The second observer's clock read –86 seconds at the start. What did it read at the end?

Exercise 2.7

Tell whether the answer to each of the following questions would be a time or a duration.

A. When will we go shopping?

B. How long will it be before the bus comes?

C. When did his term in office end?

D. How old is that building?

E. When the war ended, how long had you been overseas?

F. How much time does a job like that take?

The term *instant* needs special attention because its meaning in physics is not quite the same as it is in ordinary language. In physics, an instant is always a time, never a time interval. An instant is a *when,* not a *how long.* The phrase *at an instant* means the same sort of thing as *at 2:15.*

Uniform motion

A very important quantity used in describing uniform motion is the displacement that occurs in one unit of time. This quantity is called the *velocity* of the object and is represented by the symbol *v.*

Exercise 2.8

Write an algebraic expression for the velocity of a uniformly moving object in terms of *x*, Δx, *t*, and Δt. (Some of these quantities may not be necessary.)

Exercise 2.9

Suppose an object moves from x_i = 76 cm to x_f = 13 cm with uniform motion in 0.7 second. What is its velocity?

Note that the velocity is negative. What does this mean?

In physics, there is a distinction between the terms *velocity* and *speed*. Velocity includes the direction of motion as well as the number of centimeters traveled each second. Speed does not indicate direction; it only tells the number of centimeters traveled each second. Velocity can be negative or positive, depending on the direction of motion; speed is always positive. For example, in Exercise 2.9, the velocity is –90 cm/s and the speed is 90 cm/s. In symbols, $|v|$ = s, where v is the velocity and s is the speed.

Experiment 2.10

Ask a staff member to show you the demonstration for this experiment.

Part B: Motion with changing speed

In Part B, we study the motion of objects that are speeding up or slowing down. We find

that many of the ideas that we have developed for uniform motion can be directly applied

to nonuniform motion while others must be further developed and refined.

Section 3. Introduction to nonuniform motion

In the study of nonuniform motion, ratios of very small quantities play an important role.

We begin this section by considering physical contexts for which you may already be

familiar with interpretations of ratios of small numbers.

Sample problem

Suppose we have 0.15 mL of a solution with 0.0075 g of solute dissolved in it.
Calculation of the concentration yields 5 g/100 mL. What does the number *5* tell us
about this solution?

Sample solution

The number *5* tells us the mass of solute that *would* be dissolved in the solution *if* we
had 100 mL of the same strength solution.

This particular interpretation of a ratio of small quantities is of essential importance in
the material that follows.

Exercise 3.1

The following questions involve the interpretation of ratios of small
quantities. Try to interpret the quantities in these problems in exactly the
same way as in the preceding example; that is, try to make your
interpretation follow the same pattern as the one given.

A. A metallurgist has just produced a small bead of a new alloy of rare
metals. He uses a sensitive balance and finds the mass is 280 μg (280
micrograms, 280×10^{-6} g). By examining the bead with a microscope,
he estimates its volume to be 2.1×10^{-5} cm^3. He reports that the
density of his new alloy is 13 grams for every one cubic centimeter.

     McDermott & P.E.G., U.Wash./ *Physics by Inquiry*

A beginning student asks:

> *"How can you say that the density is 13 grams for every one cubic centimeter when there is only 2.1 × 10⁻⁵ cm³ of the material in the whole world?"*

How would you explain to the student what the number 13 g/cm³ tells us about this material and how the number can apply to such a small object? (*Note:* The word *density* is not an explanation of meaning, it is just the name of the number.)

B. Suppose a bullet travels with uniform motion a distance of 2 m in 0.005 second. If we divide the displacement by the time interval, we get 400 m/s. How do you interpret the number 400 in light of the fact that the bullet only traveled 2 m?

✔ Discuss this exercise with a staff member.

The preceding problems show that even though two quantities may be very small, their ratio still can be quite large. To interpret a ratio of small quantities, we must imagine a situation that may not really exist. What we need for an interpretation is a ratio such that the denominator is one unit (e.g., 1 cm³, 1 second). Therefore, we have to imagine a situation in which the ratio remains the same but the two quantities involved may be much larger than the original ones. For example, in part A of Exercise 3.1 we had to imagine a piece of alloy with a volume of 1 cm³ and a mass of 13 g. For part B, we had to imagine that the bullet moved uniformly for a full second and traveled 400 m.

Experiment 3.2

Obtain a fan cart and find out how to use the ticker-tape timer.

A. First, see how the fan cart moves before using the ticker tape. Let the cart start from rest (stopped), turn on the fan, then release the cart and let it travel for about 2 m.

    How would you describe the motion? Is it uniform or nonuniform?

B. Now run the motion again with the cart dragging a tape through the timer. Each student should make his or her own tape.

    Examine your tape. How can you tell from the tape whether the motion is uniform or nonuniform?

C. Divide your tape into intervals of six spaces as shown.

The timer is constructed so that it hits the tape 60 times each second. What is the duration of each six-space interval; that is, how much time does each six-space interval take? Explain your reasoning.

D. Divide the total displacement by the total duration of the trip. Can this number be interpreted as the number of centimeters traveled each second? Explain.

Measure the displacement during the first half and during the last half of the entire trip. Record your measurements in a table like the one below.

E. In our work with uniform motion, we have used the number of centimeters traveled in one second to tell us how fast an object is moving. A student might suggest:

"Let's just measure how far the object goes in each second since we can't calculate it. That would tell us the speed."

Pick a particular one-second interval and compare the displacement during the first 0.5 second and the last 0.5 second of this interval. Record the displacements in your table.

Explain why, in this case, the number of centimeters actually traveled in one second does not tell us all there is to know about the speed of the object.

F. Now focus your attention on just one of the 0.1-second (six-space) intervals. (Choose an interval near the middle of the tape.) How does the displacement in the first 0.05 second of this small interval compare with the displacement in the last 0.05 second? Record your measurements.

Interval	Displacement in first half	Displacement in second half
Whole motion	_____	_____
One-second interval	_____	_____
0.1-second interval	_____	_____

G. The most important observation to make in part F about the motion in the 0.1-second interval is that the distances traveled in the first and last halves of the interval are nearly the same.

> The measurements in part F can be considered evidence that the motion in a very small interval of time is very nearly uniform. Explain how your measurements in part F support this statement. (Make use of your operational definition of uniform motion.)

H. Calculate $\Delta x/\Delta t$ for the interval you used in part F. Give an interpretation of this number in the same pattern as in the sample problem at the beginning of Section 3 and as in Exercise 3.1.

I. Suppose we looked at a very small interval, say one millisecond. Would you expect the motion to be more or less uniform than during the 0.1-second interval? Explain.

Inadequacy of proportional reasoning for nonuniform motion

Until now, we have been using the same kind of reasoning in all of the quantitative problems we have encountered, namely proportional reasoning. In nonuniform motion, however, we have a situation where this kind of reasoning is not correct. Before we attempt to deal with nonuniform motion, let us consider carefully why the proportional reasoning that has served us so well is inadequate now.

In all of the problems we have dealt with so far, there was a constant ratio between two quantities because there was equal sharing of one quantity by equal units of another.

For example:

(1) When we studied mass and volume, we found that all of the volume units shared the mass equally if the object was homogeneous. There was the same mass for each unit of volume in all samples of a particular substance. Because of this equal sharing, the number obtained by dividing the total mass by the total volume could be interpreted as the mass of each one unit of volume. We were also able to calculate the mass of a particular volume of material from information about a different sized piece of the same material.

(2) In uniform motion, the units of time all share the distance traveled equally. The quantity $\Delta x/\Delta t$ can be interpreted as the distance traveled in each one unit of time, and it is the same number in every interval of the motion. This relationship made possible calculations such as the one in Experiment 1.4, where we used information about the motion in one interval to predict how long it would take the object to cross another interval.

In nonuniform motion, the distance traveled is not shared equally among the units of time. For this reason, many calculations that are possible for uniform motion cannot be done by the same reasoning for nonuniform motion. In Experiment 1.5, for example, an object was speeding up. We measured how long the object took to travel across one interval, but we could not use this information to predict how long the object took to cross another interval. Furthermore, there is no simple interpretation of the result of dividing Δx by Δt for a nonuniform motion like that of the fan cart. We cannot even say what the velocity of the fan cart is. It does not have a single speed: it has many speeds ranging from very low speeds at the beginning to high speeds at the end.

There are two major consequences of this lack of proportionality between Δx and Δt. For nonuniform motion:

(1) The ratio of displacement to duration in an interval of motion, $\Delta x/\Delta t$, does not have a simple physical interpretation.

(2) It is not possible to tell how far an object will move in a particular interval if we are given information about the speed in a different interval.

A new approach

We need something new in order to give a mathematical description of nonuniform motion. The key to a quantitative understanding of nonuniform motion is to be found in the results of Experiment 3.2, namely:

For very small intervals,
motion is very nearly uniform.

This important result means that we can apply what we know about uniform motion, at least in a limited way, to nonuniform motion. We cannot describe the motion as a whole, but since a small interval of the motion is nearly uniform, we can describe what happens in a small interval. In particular, we can again describe velocity mathematically, but now the interpretation of the velocity will only apply to a small interval of the motion.

Exercise 3.3

Let us consider in detail the relation of positions and times in a small interval of motion. The table below gives the position of a moving object at various times.

Time (seconds)	Position (cm)
6.72	14.8762
6.73	17.3335
6.74	19.7915
6.75	22.2502
6.76	24.7096
6.77	27.1697
6.78	29.6305
6.79	32.0920
6.80	34.5542
6.81	37.0171
6.82	39.4807

A. This motion is not perfectly uniform. How can you tell?

B. The motion is, however, very nearly uniform. Calculate $\Delta x/\Delta t$ for the entire interval shown.

C. Suppose that, instead of moving as shown in the table, the object moved with uniform motion for 0.03 second with the velocity that you calculated in part B. If it started at position 14.8762 cm, where would it be after 0.03 second?

D. Where was the object actually located at $t = 6.75$ seconds according to the table? How well does the uniform motion calculation in part C compare with the actual position?

E. Compute $\Delta x/\Delta t$ for each 0.01-second interval. How do these numbers compare with each other? How do these numbers compare with your result in part B?

F. Explain why we can say that the motion in the 0.01-second interval is very nearly uniform.

Definition of instantaneous velocity

On the basis of the experiments and analysis that we have conducted thus far, we now introduce a new concept: *velocity at an instant.* If we want to know the velocity of an object at any particular instant, what we need to do is look at a very small interval that contains that instant. It is convenient to have the instant of interest be at the middle of the small interval. We must first make sure the interval is small enough that the motion is nearly uniform. We then measure Δx, measure Δt, and divide. The number obtained this way is called the *instantaneous velocity* at the instant of interest. It will be represented by the symbol v. The magnitude of the instantaneous velocity is called the *instantaneous speed*. From now on, the terms *velocity* and *speed* will always be understood to mean instantaneous velocity and instantaneous speed, respectively.

Interpretation of instantaneous velocity

Recall that for intervals of uniform motion less than one second, a velocity of 400 m/s had this interpretation: The object *would* travel 400 m *if* it continued to move in the same way for an entire second. In the case of instantaneous velocity, we have a similar interpretation, but now there is another *if* involved.

Suppose we obtain 45 cm/s for $\Delta x/\Delta t$ in a small interval. We make the interpretation that the object *would* travel 45 cm *if* it continued to move at the same velocity (without speeding up or slowing down) and *if* the motion continued that way for an entire second.

How small an interval?

How small should our intervals be? How nearly uniform is adequate for our definition of instantaneous velocity? These questions are surprisingly subtle. To do a good job answering them would require a long digression into the mathematics of limits. Since this topic is usually covered in calculus courses, we will not give a careful presentation here.

Remember, the interval should be small enough that the motion is nearly uniform. The criterion for uniformity may be different in different circumstances. In one case for a ticker tape we might be satisfied if the spaces between the dots were the same to within 10%. In another case, we might want the motion to be so nearly uniform that we could not see any difference in the dot spacing just by looking. Another time we might need the dot spacing to be the same to within the uncertainty of our measuring instruments.

Exercise 3.4

Suppose we are looking at an interval of a motion that is nearly uniform. Now suppose we look at a small segment of that interval, say one-tenth of it. Would you expect that motion in this smaller interval would also be nearly uniform? Would you expect the motion to seem more or less uniform in the smaller interval? Explain your reasoning.

Use of instantaneous velocity to compare speeds

In the following exercise, we will use the concept of instantaneous velocity to help us recognize how to compare the speeds of two object. In order to apply the concept of instantaneous velocity, we must focus attention on small intervals of motion. The motion must be considered bit by bit rather than as a whole.

Exercise 3.5

A. *One object faster than another*

Suppose that at a given instant one object is moving faster than another. The number obtained for the instantaneous speed of the faster object will be larger than that obtained for the instantaneous speed of the slower one. (Statements about instantaneous speed always apply to just one instant, since things might be different at another time.)

Example: Object B is faster than object A.

Let us think about a small interval of time and see what happens to each object during that time interval. Suppose that object A has an instantaneous velocity of 200 cm/s and object B has an instantaneous velocity of 300 cm/s at the time we are comparing them. Suppose object B is ahead of object A at this time.

How far does each object move in an interval of 0.01 second?

If we made a diagram similar to those in Section 2, we would have:

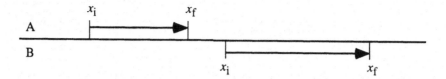

How can you tell from this diagram that object B is faster?

Measure the distance between object A and object B at the beginning of the interval and again at the end of the interval. How is the distance between the objects changing?

Draw a similar diagram for the case in which the faster object is behind the slower one.

How is the distance between the objects changing in this case?

B. *Two objects with the same velocity*

Now suppose the two objects have the same velocity; that is, they have the same instantaneous velocity at a particular time. Suppose one object is ahead of the other.

Make a diagram as in part A to illustrate this situation.

How is the equality of the instantaneous velocities shown on the diagram?

How is the distance between the two objects changing in this case?

C. *One object passing another*

Consider two objects traveling in the same direction. We will analyze their passing in two stages: just before passing and just after.

First, look at the situation just before passing; that is, look at a time interval in which the objects are side by side at the end of the interval. Assume object B is passing object A. That means that just before the actual passing, object B was behind object A.

Draw a diagram indicating the position of object A and object B at the beginning and end of the interval, as you have done in parts A and B.

Which object travels farther in the time interval?

Which object had the larger instantaneous velocity?

Now look at a time interval that begins with the objects side by side. This interval begins at the same time as the previous interval ends.

If object B is passing object A, which is ahead at the end of the interval?

Which has the larger instantaneous velocity?

Illustrate the motion in this interval with a diagram.

If one object passes another, is it possible for them to have the same instantaneous velocity at that instant?

Is it possible for two objects to be side by side and have the same instantaneous velocity? Illustrate your answer with a diagram.

✔ Check your results with a staff member.

Experiment 3.6

Ask a staff member to run the demonstration for you. Apply the concept of instantaneous velocity to the demonstration.

Exercise 3.7

Give an operational definition of instantaneous velocity in your own words.

Exercise 3.8

Consider the following statement by a student:

"$\Delta x/\Delta t$ gives the velocity for an interval. To find the velocity at an instant, divide the position at that instant by the time at that instant: $v = x/t$."

Does this statement express the same ideas as those in the preceding text or does it express different ideas? Is $v = x/t$ a precise mathematical statement of the definition of instantaneous velocity? Explain.

Sample problem

Wheat is exported from many countries. The wheat export rate for a country tells how fast wheat is leaving the country. An operational definition of wheat export rate is:

$$\frac{\text{Mass of wheat exported during a certain amount of time}}{\text{That amount of time}}$$

This rate varies throughout the year and from year to year. Make a mathematical analogy between wheat export rate and velocity. Tell what corresponds to instantaneous velocity, displacement, and duration.

Sample solution

Wheat export rate corresponds to instantaneous velocity. Both are ratios (numbers obtained by division). Both quantities are not necessarily constant. To compute both, we must consider a small enough interval that the rate is nearly constant. (We cannot calculate the export rate using an interval such as one year if we want information about the high and low rates during the year.)

The amount of wheat exported in an interval of time corresponds to displacement. Both are in the numerator of the ratio.

Duration of motion corresponds to duration of wheat exportation. Duration is in the denominator in both ratios.

Exercise 3.9

One observation commonly taken during a physical examination is the patient's pulse rate before, during, and after vigorous exercise.

A. How would you expect the pulse rate to vary in such an examination?

B. Give an operational definition of pulse rate. Your definition should work when the time of measurement is of any short duration of time, not a special amount of time like one minute or 10 seconds.

C. Make a mathematical analogy between pulse rate and velocity. Identify what corresponds to instantaneous velocity, displacement, and duration, and tell how corresponding quantities are alike.

Section 4. Changing velocity

Exercise 4.1

A ball rolls from point *A* to point *D* on the tracks shown below. The length of each track and the time the ball takes to roll across that section of track is indicated.

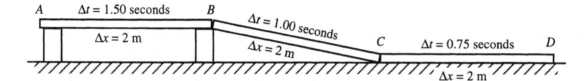

Consider the following statement by a student:

> *"If I want the velocity of the ball on a segment of track, I just take that Δx and divide by that Δt. Velocity is just Δx/Δt."*

A. Do you agree with the statement above when it is applied to the first segment of track *AB?* Explain your reasoning.

Calculate the quantity $\dfrac{\Delta x}{\Delta t}$ for the first segment of track.

Does that quantity represent the velocity of the ball at the start of the first track? Does that quantity represent the velocity of the ball at the end of the first track? Explain.

B. Do you agree with the statement above when it is applied to the second segment of track *BC?* Explain your reasoning.

Calculate the quantity $\dfrac{\Delta x}{\Delta t}$ for the second segment of track.

Does that quantity represent the velocity of the ball at the start of the second track? Does that quantity represent the velocity of the ball at the end of the second track? Explain.

C. Do you agree with the statement above when it is applied to the third segment of track *CD?* Explain your reasoning.

Calculate the quantity $\dfrac{\Delta x}{\Delta t}$ for the third segment of track.

Does that quantity represent the velocity of the ball at the start of the third track? Does that quantity represent the velocity of the ball at the end of the third track? Explain.

Exercise 4.2

Several students do an experiment involving three balls and tracks. The first track is inclined as shown. The second and third tracks have been arranged to give uniform motion. The balls on the second and third tracks are released from specific marks on the launching ramps so that the motions can be replicated as many times as needed.

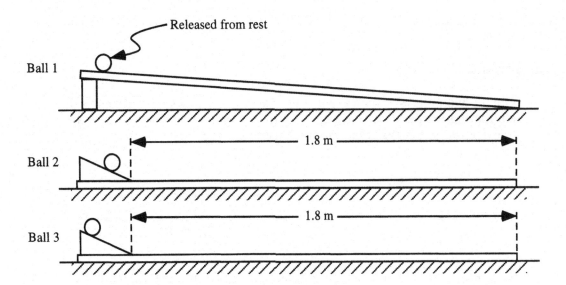

The students have collected the following data:

Ball 2 takes 2.3 seconds to travel 1.8 m.

Ball 3 takes 1.5 seconds to travel 1.8 m.

Ball 1 and ball 2 have the same speed 0.8 second after ball 1 is released.

Ball 1 and ball 3 have the same speed 1.2 seconds after ball 1 is released.

A. Find the speed of ball 1 at a time 0.8 second after it is released and also at a time 1.2 seconds after it is released. Explain your reasoning.

B. Suppose the students wished to find the times at which ball 1 had other speeds as well. Describe in detail an experiment the students could conduct to determine the speed of ball 1 at other instants.

✔ Check your answers with a staff member.

McDermott & P.E.G., U.Wash./*Physics by Inquiry*

A ball rolling on an incline is continually changing its speed, thus we cannot directly apply our operational definition of velocity. However, we can find the speed of the ball at an instant by comparing its motion to that of an object that is moving with uniform speed. (See the preceding experiment and Experiment 2.10.) In the next experiment we use a uniformly moving belt to investigate how the speed of a ball on an incline changes.

Experiment 4.3

Obtain two pulleys, a motor, and a long string or belt that is marked with alternating light and dark stripes. The motor should have several settings to drive the belt at different, constant speeds.

The belt provides a constant speed that you can compare with the motion of a ball rolling on a track. By simultaneously watching the ball and the belt, you can identify the instant at which the two have the same speed.

A. Arrange the belt so that it moves parallel to the table. Set up tracks next to the belt as shown so that the first half of the track slopes downward and the second half is level.

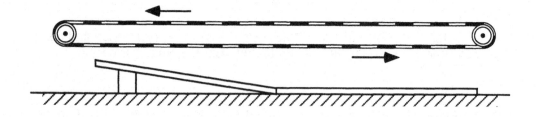

Set the belt at an intermediate speed. Release the ball from various places on the inclined section.

Where on the incline can you release the ball so that when it reaches the level track it has the same speed as the belt? Demonstrate this motion to a staff member before continuing.

B. Set up the tracks so that they form a single, continuous inclined track between 2.5 and 3 meters long. Incline the belt as shown.

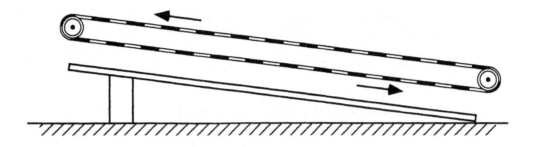

Set the belt on a low speed. Release the ball from rest and measure how much time is required for the ball to reach the belt's speed.

Repeat this experiment near the top of the track, in the middle of the track, and near the bottom of the track.

What do you conclude from these measurements?

C. In this part of the experiment we will find the speed of the ball at several instants during the motion. Since the speed is changing, it is difficult to make a direct measurement of the speed at an instant. We can, however, determine the instant when the ball has the same speed as the belt. We then make a separate measurement to find the belt speed. We can then say: "The ball had a velocity of $v =$ ___ at time $t =$ ___ ." Using the belt is an indirect way of finding the speed of the ball at a particular instant.

(1) Before you take measurements, check the incline of the track. The ball should reach the fastest belt speed near the bottom of the track. Change the angle of the track if necessary.

(2) Set the motor to the lowest speed. Release the ball from rest at the top of the track and measure the time required for the ball to reach the belt speed. The person who releases the ball should start the clock at the same time. Also measure the belt speed.

Note: When the belt speed is changed, it may not be possible to recover the same belt speed exactly. Therefore *all* necessary measurements should be taken before the speed is changed.

(3) Repeat part 2 above for at least three more belt speeds.

Is it necessary to start the ball from the same location on the track each time? Explain your reasoning.

After completing the measurements, you should know when the ball had each of the belt speeds on its way down the track. This information is a kind of history of the ball's motion.

D. Plot a velocity versus time graph for the ball's motion.

What does just one point on the graph tell you?

Is the origin, *(0,0),* a real, observed data point? Explain your reasoning. Should the origin be included on the graph?

What can you conclude about the motion from the straightness of the graph?

Calculate and interpret the slope of the graph.

✔ Discuss this experiment with a staff member.

Section 5. Acceleration

Exercise 5.1

A car travels with constant velocity 20 mph (miles per hour) until 3:50:00 and then speeds up as shown on the graph below. The driver notices that it takes 3 seconds to speed up from 20 mph to 50 mph.

In answering the following questions, explain your reasoning.

A. When does the car have a velocity of 75 mph?

B. How fast is the car going at 3:50:02?

C. Can you determine how far the car traveled between 3:49:55 and 3:50:00?

D. Can you determine how far the car traveled between 3:50:00 and 3:50:05?

E. Calculate the slope of the inclined section of the graph below.

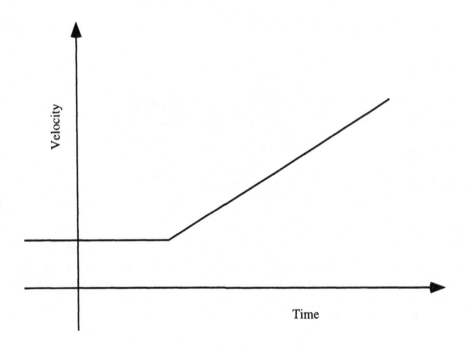

The number calculated for the slope of the graph in Exercise 5.1 is called the

acceleration. Similarly, the slope of the graph in Experiment 4.3 is the acceleration of the

ball rolling down the track. Both of these motions have the characteristic that the velocity

of the object changed by the same amount in equal time intervals. When an object's

motion has this characteristic, we say that the object has *constant acceleration.* In this

case, the total change in velocity is shared equally by all equal time intervals. We can

therefore interpret the number $\Delta v/\Delta t$ as the *change* in velocity occurring in each unit of

time. The number, $\Delta v/\Delta t$, is called the *acceleration* and is represented by the symbol *a*.

$$\text{\textit{Symbolically,} } a \ = \ \frac{\Delta v}{\Delta t}, \text{ \textit{if acceleration is constant.}}$$

In this section, we only deal with situations in which the acceleration is constant.

Acceleration can mean speeding up or slowing down: any change in velocity qualifies.

Sample problem

Suppose an object is speeding up with constant acceleration. At $t = 0$, the object has
velocity 4.6 mph. At $t = 90$ seconds, the velocity is 8.2 mph.

    A. What is the acceleration?

    B. What is the velocity at $t = 15$ minutes?

Sample solution

    A. From $t = 0$ to $t = 90$ seconds, the change in velocity is $8.2 - 4.6 = 3.6$ mph.
       The duration of this interval is 90 seconds. The acceleration is the change
       in velocity, Δv, divided by the duration of the interval, Δt, so the
       acceleration is 3.6 mph/90 seconds = 0.04 mph/s.

    B. Since the acceleration is constant, the change in velocity is 0.04 mph
       in each second. From $t = 0$ to $t = 15$ minutes, 900 seconds elapse. We
       therefore need to add 0.04 mph 900 times, once for each second, in
       order to find the total change in velocity from $t = 0$ to $t = 15$ minutes.

       This addition can be done quickly by multiplying 900×0.04, which
       gives 36 mph. This is the change in velocity from $t = 0$ to $t = 15$ minutes.
       The velocity at $t = 0$ is 4.6 mph, so the velocity at $t = 15$ minutes must be
       $4.6 + 36 = 40.6$ mph.

Exercise 5.2

Consider the following situation: A boat is speeding up with constant acceleration. The boat speeds up from 10 mph to 22 mph in 3 seconds. A student who is studying this motion subtracts 22 – 10 and obtains 12.

How would you interpret the number *12?*

Exercise 5.3

An object moving with constant acceleration has a velocity of 6 mph at $t = 0$ and a velocity of 30 mph at $t = 6$ seconds.

Find the velocity at $t = 4$ seconds. Explain your reasoning. Do not use algebra. (That is, do not base your answer solely on an algebraic formula.)

Exercise 5.4

An object has a constant acceleration of 50 mph/s. After accelerating for a time *T,* its velocity is 160 mph. (*Note: T* is a duration.) Write an algebraic expression for the initial velocity in terms of *T.*

In our first examples of acceleration, we have worked with velocities expressed in miles per hour to make it easier to give an interpretation of the number obtained for the acceleration. But metric units are almost always used in the sciences. An acceleration of 5 mph per second would be 2.2 m/s per second in the metric system since 5 mph is 2.2 m/s. This acceleration could be written 2.2 m/s/s, read "2.2 meters per second per second." In practice, this is often written 2.2 m/s^2, read "2.2 meters per second squared." A square second makes no sense; m/s^2 is just a shorthand way of writing m/s/s.

It is important to realize that the two *seconds* in m/s/s play different roles. The *second* in m/s is just there as part of the velocity unit (like *hour* is part of the unit *mph*). The other *second* is the unit of time we use when telling how much the velocity changes in one unit of time.

Exercise 5.5

A car is speeding up with constant acceleration. If the car speeds up from 10 m/s to 34 m/s in 5 seconds, how much longer will it take the car to reach a speed of 53 m/s? Explain your reasoning. Do not use algebra.

Exercise 5.6

Tell what is wrong with each of these statements.

A. *"Acceleration tells how fast the object is moving per second."*

B. *"Acceleration tells how much the velocity increases."*

McDermott & P.E.G., U.Wash./ *Physics by Inquiry*

Part C: Graphical representations of motion

In Part A and Part B of *Kinematics,* we have developed the concepts of position, displacement, velocity, and acceleration. In Part C, we examine how to represent motion graphically and how to interpret motion graphs.

Section 6. Motion and graphs

Exercise 6.1

Answer the following questions about the velocity versus time graph below. Explain your reasoning in each case.

A. What does the single point *A* indicate about the motion?

B. Give an interpretation of the length labeled *c*.

C. Give an interpretation of the length labeled *d*.

D. Give an interpretation of the ratio *d/c*.

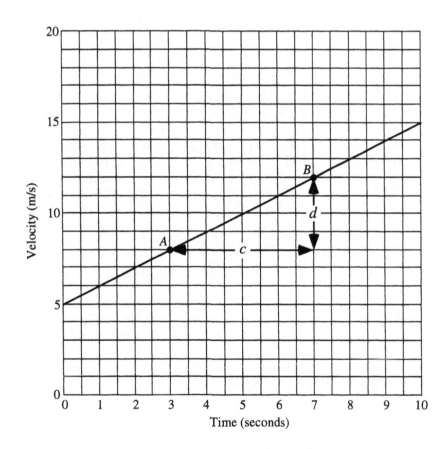

Experiment 6.2

Set up two boards as shown and obtain a skate car and ticker-tape timer.

Brick   Brick

A. Without using the ticker tape, release the car at the top of one of the boards and let it move until it starts to roll back.

> How do you think the instantaneous velocity varies throughout the motion? Answer by thinking about what happens during small intervals during the motion.

B. Make a ticker-tape record of this motion. Divide your tape into 0.1-second intervals. Is the motion nearly uniform in each interval?

C. Review the definition of position in Section 2. Make a position versus time *(x* vs *t)* graph for this motion. Plot points every 0.1 second.

> Identify the part of the graph that corresponds to the car reaching the bottom. Explain your reasoning.

D. Make a table of times and corresponding velocities for the motion. Make an entry every 0.1 second.

> When you calculate an instantaneous velocity for a small interval, you must decide which instant it describes. Should the value for the instantaneous velocity be associated with the time at the beginning of the interval, at the end, or some time in the middle? Explain your reasoning.

E. Plot a graph of instantaneous velocity versus time *(v* vs *t)* for this motion.

> How do you interpret your graph? Identify the different segments of the graph with the corresponding parts of the actual motion.

> Compute the slopes of the straight sections of your graph. How do you interpret these numbers?

✔ Discuss your graphs with a staff member.

Qualitative interpretation of curved position versus time graphs

For straight-line position versus time graphs, the velocity of the uniform motion is given by the slope of the graph. This relationship between the velocity and slope also holds for curved graphs such as the one in Experiment 6.2.

Sample problem

Describe the variation of the speed of the motion represented on the following graph.

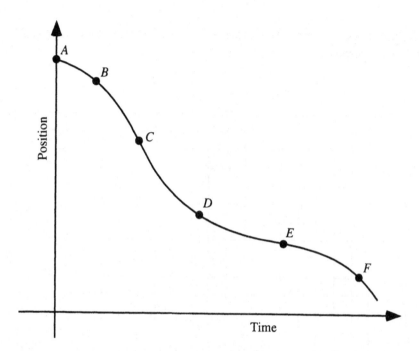

Sample solution

At point *A*, the earliest instant we know about, the graph has a downward, or negative, slope. The object's initial velocity is therefore negative; that is, the object is moving in the negative direction.

At point *B*, the curve is getting steeper, so the object is speeding up.

At point *C*, the curve is steepest (steeper than at *B* or *D*), so the object's speed is highest at this point.

At point *D*, the curve is getting less steep or flatter, so the object is slowing down. It is still moving in the same (negative) direction.

At point *E*, the curve is flattest, so the motion is slowest.

At point *F*, the curve is again getting steeper, so the object is speeding up.

McDermott & P.E.G., U.Wash./*Physics by Inquiry*

Exercise 6.3

Answer the following questions by carefully examining the graph below.
Explain your reasoning in each case.

A. During approximately what interval of time (i.e., from $t = ?$ to $t = ?$) is
the object speeding up?

B. During approximately what interval of time is the object slowing
down?

C. During approximately what interval of time is the object moving with
constant speed?

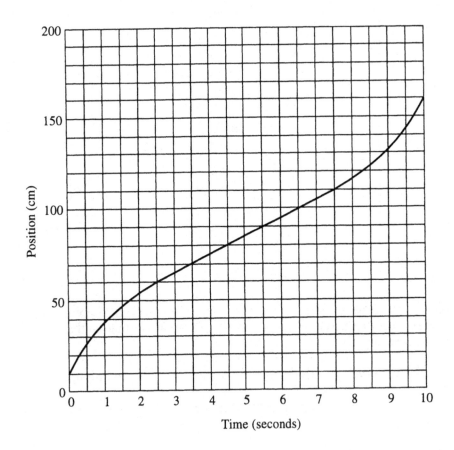

McDermott & P.E.G., U.Wash./ *Physics by Inquiry*

Exercise 6.4

At which of the lettered points on the graph below:

- is the motion slowest?

- is the object speeding up?

- is the object slowing down?

- is the object turning around?

Explain your reasoning in each case.

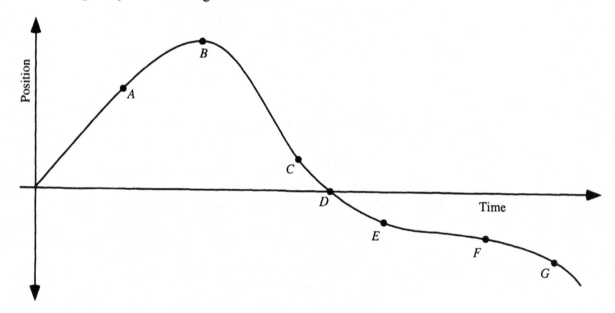

          McDermott & P.E.G., U.Wash./ *Physics by Inquiry*

Exercise 6.5

Answer the following questions by carefully examining the graph below. Explain your reasoning in each case.

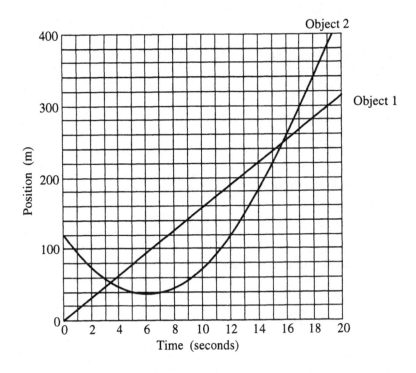

A. Give an example of a time when the instantaneous velocity of one of the objects is zero.

B. Which object's speed is larger at $t = 13$ sec?

C. Which object's speed is larger at $t = 18$ sec?

D. At what time do the objects have the same velocity?

E. How far apart are the objects when they have the same velocity?

✔ Check your reasoning with a staff member.

McDermott & P.E.G., U.Wash./ *Physics by Inquiry*

Section 7. Curved graphs

We can apply many of the ideas developed to describe nonuniform motion to the interpretation of curved graphs. If a graph is curved, it does not rise by the same amount for each unit of run. Therefore, the ratio *(total rise)/(total run)* has no simple interpretation. This situation is very similar to the one we faced in nonuniform motion, and the methods we used there can also be used here.

Exercise 7.1

A. Below is a position versus time graph of a nonuniform motion.

Would you describe the motion as a whole (from $t = 0$ to $t = 8$ seconds) as nearly uniform or as definitely nonuniform? Explain your reasoning.

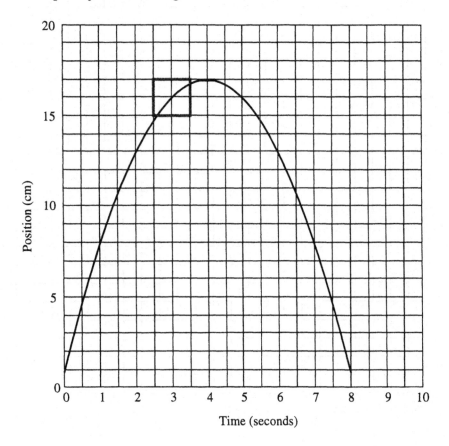

We analyze the graph in part A in detail around $t = 3$ seconds and $x = 16$ cm by creating an expanded view of the boxed section.

McDermott & P.E.G., U.Wash./ *Physics by Inquiry*

B. Consider just the portion of the graph from $t = 2.5$ seconds to $t = 3.5$ seconds. The position and time coordinates for points in this interval are given below. Plot these points on the graph provided to create an expanded view of this small interval.

t (seconds)	x (cm)	t (seconds)	x (cm)
2.5	14.75	3.1	16.19
2.6	15.04	3.2	16.36
2.7	15.31	3.3	16.51
2.8	15.56	3.4	16.64
2.9	15.79	3.5	16.75
3.0	16.00		

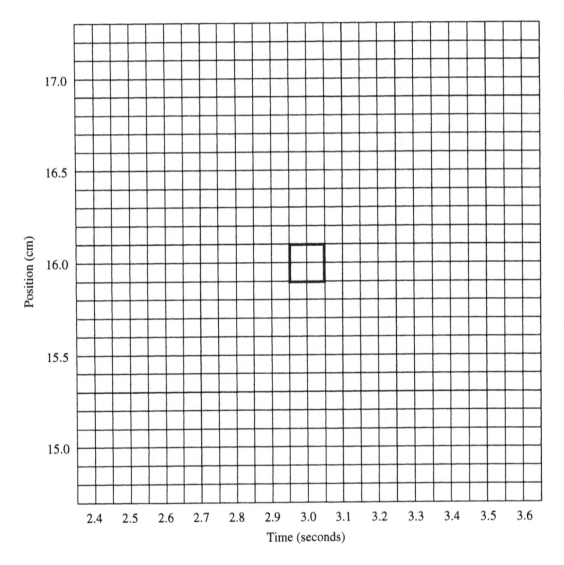

Next we expand the section of this graph in the small box at the center.

C. Position and time coordinates are given below for points in the interval from $t = 2.95$ seconds to $t = 3.05$ seconds (the section in the very small box at the center of the preceding graph). Plot these points on the graph provided.

t (seconds)	x (cm)	t (seconds)	x (cm)
2.95	15.897	3.01	16.020
2.96	15.918	3.02	16.040
2.97	15.939	3.03	16.059
2.98	15.960	3.04	16.078
2.99	15.980	3.05	16.098
3.00	16.000		

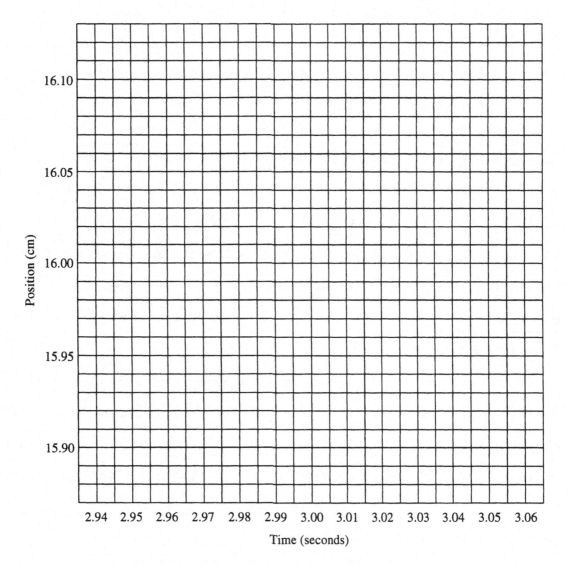

All three graphs are representations of the same motion. Why is the last one so much straighter?

✔ Discuss this exercise with a staff member.

Slope of a curved graph

With straight line graphs, we discovered that the slope can tell us a great deal about the physical situation. It tells how changes in one quantity are related to changes in another. With a curved graph, this relation is always varying and must be determined separately for each point of interest.

If we want to find out how much the graph is rising per unit of run at a particular point, we must look at a small interval that contains the point. If the interval is small enough, the graph behaves much like a straight-line graph. We can then find the slope just as with a straight-line graph: the rise divided by the run. The number obtained this way is called the *slope at a point,* or just the *slope.* Of course, this number is valid only within the small interval.

Finding the slope of a curved graph: tangents

Recall that for instantaneous velocity, we made the following interpretation: This number tells how far the object *would* travel *if* it traveled at the same speed (without speeding up or slowing down) and *if* the motion continued that way for one entire unit of time. A similar interpretation can be made for the slope of a curved graph. In fact, the interpretation may be easier in the case of graphs because we can illustrate it by drawing a line on the graph.

We interpret the slope as the *rise per unit run* the graph *would* have *if* the graph did not curve anymore but continued as a straight line. This imaginary straight line is called the *tangent to the graph* at the point of interest. To draw the tangent, we imagine looking at a very small interval around the point of interest. This small section of the graph will be very nearly straight. We then extend this straight line in both directions to form the tangent.

Sample problem

Find the slope of the following graph at $t = 3$ seconds.

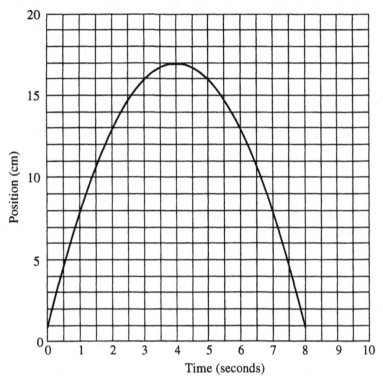

Sample solution

First draw the tangent at $t = 3$ seconds, as shown.

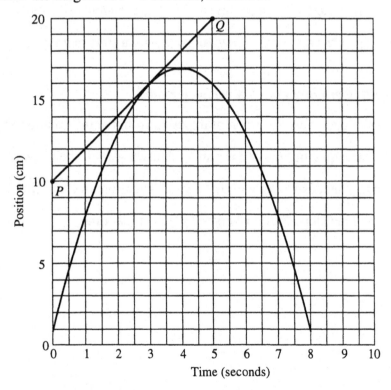

We can now evaluate the slope of the graph at $t = 3$ seconds by calculating the slope of the tangent, which is just an extension of the nearly straight line in a small interval around the point at $t = 3$ seconds. We can then find the slope of the tangent by using any two points on it. Choosing two points far apart minimizes errors due to graph reading. Using points P and Q, we get $(20 - 10)/(5 - 0) = 2$ cm/s for the slope at $t = 3$ seconds. This number is the instantaneous velocity at $t = 3$ seconds.

The slope of a position versus time graph at an instant

gives the instantaneous velocity at that instant.

Exercise 7.2

Calculate the slope of the last of the three graphs in Exercise 7.1. How is this graph relevant to the preceding sample problem? How do the slopes compare?

Always calculate the slope of a curved graph by drawing the tangent, extending it a great distance on both sides of the graph point, and choosing two points on the tangent that are far apart. It is a bad practice to evaluate the slope of a graph by choosing two points very close together along the graph because errors in reading the graph can make the number obtained practically meaningless. Drawing tangents is a skill that takes practice to develop. Check your first few tangents with a staff member.

Plotting a v *versus* t *graph given an* x *versus* t *graph*

It is possible to plot the v versus t graph of a motion from the information available on the x versus t graph. The quantity $\Delta x/\Delta t$ for a small interval is *both* the slope of the x versus t graph and the instantaneous velocity of the object. Therefore, we can find the velocity at any time by finding the slope of the x versus t graph at that time.

Exercise 7.3

A. Construct the *v* versus *t* graph for the motion represented on the following graph. *Use graph paper!* Plot points every one second.

B. For both graphs, identify the parts that correspond to speeding up.

   (1) How can you tell the object is speeding up from looking at the *x* versus *t* graph?

   (2) How can you tell the object is speeding up from looking at the *v* versus *t* graph?

C. For both graphs, identify the parts that correspond to slowing down.

   (1) How can you tell from the *x* versus *t* graph?

   (2) How can you tell from the *v* versus *t* graph?

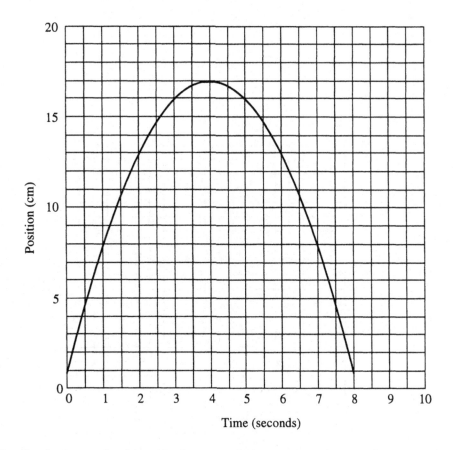

D. For both graphs, identify the parts that correspond to turning around.

   (1) How can you tell from the *x* versus *t* graph?

   (2) How can you tell from the *v* versus *t* graph?

E. Find the acceleration at $t = 2$ seconds, $t = 4$ seconds, and $t = 6$ seconds.

F. Fill in the following table:

t	v	a	slowing down or speeding up?
2 seconds			
4 seconds			
6 seconds			

✔ Check your results with a staff member.

Kinks on a graph

So far we have not discussed graphs with corners or kinks in them. A graph with a kink is shown below.

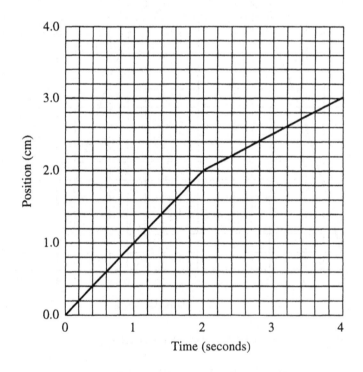

The distinguishing characteristic of a kink is that the graph has an abrupt change of slope at that point. In the preceding graph, for example, the slope is 1 m/s at every instant from $t = 0$ up until $t = 2$ seconds. From $t = 2$ seconds to $t = 4$ seconds, however, the slope is 0.5 m/s all the time. Right at $t = 2$ seconds, it is hard to say what the slope is, but we can say that the slope is experiencing an abrupt change. If we were to plot a v versus t graph for this motion, there would be two sections of constant velocity, with the change coming at $t = 2$ seconds. There are three ways in which this behavior is customarily represented on a v versus t graph:

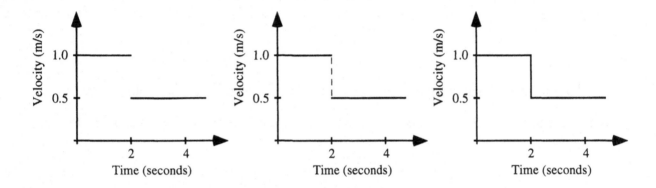

Kinks on curved graphs mean exactly the same thing as on straight-line graphs—an abrupt change in slope. With curved graphs, the slope is always changing. However, for a smooth curve, the slope changes gradually, whereas at a kink, the slope changes abruptly. In the following graph, the slope increases gradually until $t = 2$ seconds. Then the slope abruptly changes sign and becomes a negative number, gradually returning to zero.

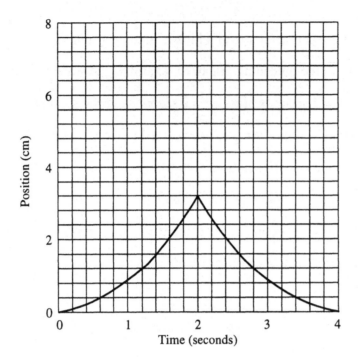

To see what happens to the slope, we examine the tangents as shown below. The slope just before $t = 2$ seconds is $+3$ m/s, and the slope just after $t = 2$ seconds is -3 m/s.

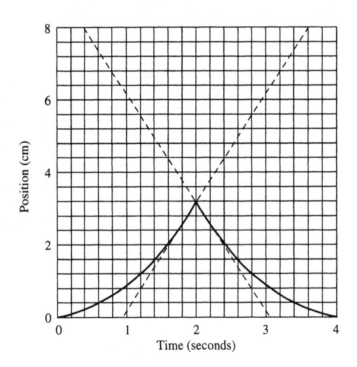

The velocity versus time graph that corresponds to the preceding graph would look something like the following graph, with gradual changes in the velocity except for a jump at $t = 2$ seconds.

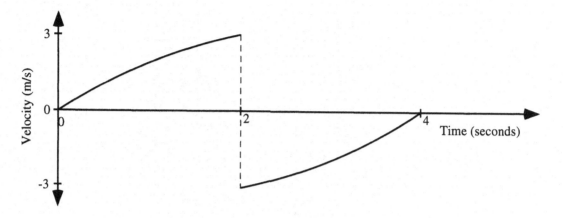

Exercise 7.4

Sketch the v versus t graph that corresponds to the x versus t graph below. *Sketch* means draw the general shape of the graph without computing actual values. Label t_0, t_1, t_2, t_3, and t_4 on your graph.

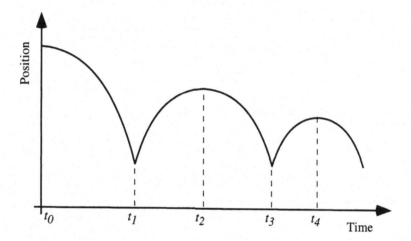

Comparing graphs and their slopes

Many situations arise in which we must compare the heights, or vertical coordinates, of two graphs. There are also situations in which we must compare the slopes of graphs. It is always necessary to decide whether the desired information is represented by the height of the graph, or the slope, or neither.

McDermott & P.E.G., U.Wash./ *Physics by Inquiry*

Exercise 7.5

Answer the following questions based on information given in the graph below. Explain your reasoning.

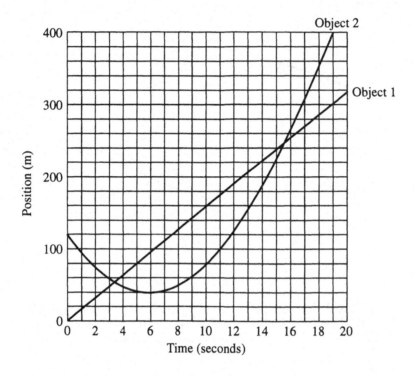

A. What are the objects' speeds at $t = 2$ seconds?

B. What are the objects' speeds at $t = 13$ seconds?

C. What are the objects' speeds at $t = 18$ seconds?

D. At what time do the objects have the same velocity and what is that velocity?

The same information represented on different graphs

A position versus time graph and a velocity versus time graph each tells much about a motion. The table you will complete in the following exercise summarizes how different information about a motion is represented on each graph.

McDermott & P.E.G., U.Wash./ *Physics by Inquiry*

Exercise 7.6

Fill in the missing entries in the table below. Before completing the table, examine the sample entries and be sure you understand them. Note that some aspects of the motion may be displayed on only one of the graphs.

How to get information from motion graphs

Information sought	*x* versus *t* graph	*v* versus *t* graph
where the object is at a particular instant		
the object's velocity at an instant		
the object's acceleration (if constant)	can't tell	compute the slope
whether the motion is uniform		
whether the object is speeding up	check whether the curve is getting steeper	
whether the object is slowing down		
whether the acceleration is constant		

✔ When you have completed your table, discuss it with a staff member.

  McDermott & P.E.G., U.Wash./*Physics by Inquiry*

Section 8. Graphs and actual motions

The skills discussed in the preceding section are powerful tools for interpretation of data and analysis of quantitative relations. The methods used are sometimes complex and a bit subtle. Occasionally there is the danger of being distracted from the reality represented and concentrating on the graph for its own sake. In this section are experiments that provide practice in maintaining the connection between the graph and the actual motion.

Experiment 8.1

Have a staff member help you run the motion for this experiment.

A. How does the velocity vary on the level sections of track?

B. On the inclined section, the velocity of the ball increases with constant acceleration. Identify and describe the experiment in which this constancy was determined.

C. Plot a *v* versus *t* graph for this motion.

Experiment 8.2

Design and set up an arrangement of tracks that will produce a motion represented on the following graph. Make all of your slopes very gradual.

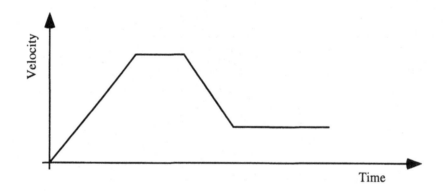

✔ Demonstrate the motion for a staff member.

Experiment 8.3

Have a staff member help you run the motion for this experiment. Plot a
v versus *t* graph for this motion.

Experiment 8.4

Design and set up an arrangement of tracks that will produce a motion
represented on the following graph.

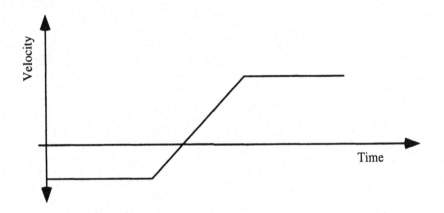

✔ Demonstrate the motion for a staff member.

Exercise 8.5

A. Sketch an *x* versus *t* graph that corresponds to the motion in
   Experiment 8.3.

B. Sketch an *x* versus *t* graph that corresponds to the motion in
   Experiment 8.4.

✔ Check your results with a staff member.

Experiment 8.6

A. Design and set up arrangements of tracks that will produce the motions represented on the two graphs.

B. Sketch the corresponding *v* versus *t* graphs for each of the motions.

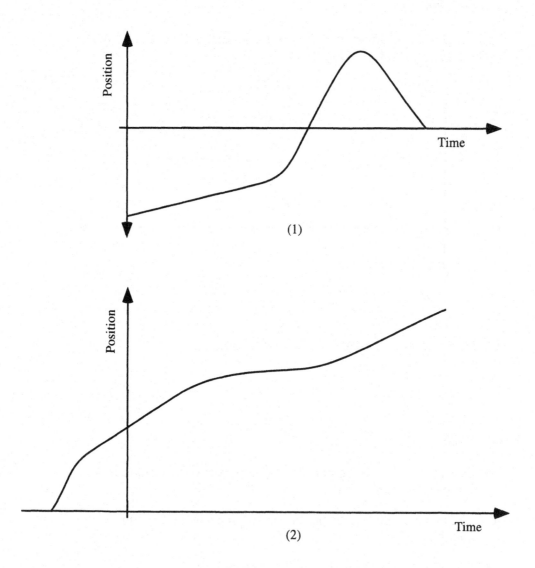

(1)

(2)

✔ Demonstrate the motions for a staff member.

Exercise 8.7

Act out the motions represented on the following graphs by sliding your finger along the table. Practice until you can do this without any pauses that are not shown on the graph.

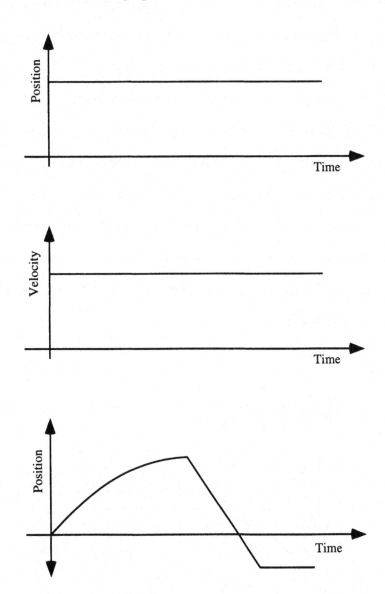

✔ When you are able to execute the motions and explain how you knew to move as you did, check with a staff member.

Section 9. Rates and graphs

The term *rate* is used in many ways. There are interest rates, gasoline mileage rates, literacy rates, electricity rates, and so on. In most cases, a rate is a number obtained by division. For example, gasoline mileage rate is the distance traveled divided by the amount of fuel used; shipping rate is the shipping cost divided by the weight of the parcel being shipped.

Many rates tell how much some quantity changes per unit of time. If we represent the quantity by q, then $\Delta q/\Delta t$ is called *the time rate of change of q* or *the rate of change of q*. Usually, rates are not constant but vary with time. If we look at a small enough interval, however, most rates are nearly constant. We can then use the concept of instantaneous rate, just as we did with velocity. In fact, velocity is called the rate of change of position.

Many of the methods developed to deal with velocity can be used in other contexts. The following sample problem illustrates the application of the concept of slope to a graph from a different context.

Sample problem

The weight of some animals varies from season to season. Below is a graph of the *weight* of a hypothetical animal versus *time*.

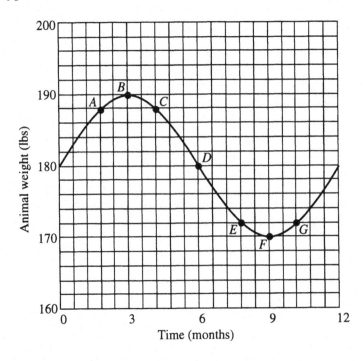

        McDermott & P.E.G., U.Wash./*Physics by Inquiry*

The *weight change rate* in this problem is the rate of change of the animal's weight. Describe the behavior of the weight and the weight change rate at points *A, B,* and *C.*

Sample solution

In solving this problem, we must remember that height represents weight, and slope represents weight change rate.

Point *A:*   The graph is heading upward, so the weight is increasing and the weight change rate is positive. The slope is getting smaller, so the weight change rate is decreasing. Since the slope is positive, this means the animal's weight is increasing less quickly.

Point *B:*   The graph is at its highest point, so the weight is at a maximum. The slope here is 0, so the weight change rate is 0; the weight is in between increasing (earlier) and decreasing (later).

Point *C:*   The graph is heading downward, so the weight is decreasing and the weight change rate is negative. Since the curve is getting steeper, the animal's weight is changing more quickly than before. In this case, since the weight change rate is negative, the animal is losing weight more quickly than before.

Exercise 9.1

Answer the following questions about the graph in the preceding example.

A. List all of the lettered points, if any, at which each of the following is true. Explain your reasoning.

    (1)   The weight is increasing.

    (2)   The weight is decreasing.

    (3)   The weight is minimum.

    (4)   The weight change rate is zero

    (5)   The weight change rate is decreasing but the weight is increasing.

    (6)   The weight is changing fastest.

    (7)   The weight change rate is increasing but the weight is decreasing.

    (8)   The weight change rate and the weight are both decreasing.

    (9)   The weight is increasing more and more slowly.

    (10)   The weight is decreasing more and more quickly.

B. Sketch the *weight change rate* versus *time* graph.

   ✔   Check your reasoning with a staff member.

Derivatives

The process of finding the slope of a graph at a point is called *differentiation* or *taking the derivative*. The slope of a graph is called a *derivative*. The slope of a position versus time graph, for example, is called *the derivative of the position with respect to time*. The following terms all mean the same thing: instantaneous velocity, rate of change of position, and derivative of position with respect to time.

In the preceding example problem, the slope is called the derivative of the weight with respect to time, or the time derivative of the weight, or simply the derivative. Every instantaneous rate is a derivative whether or not it is obtained from the slope of a graph.

In this module, we will be concerned with derivatives only as they apply to graphs. A derivative, however, has a more general definition than just the slope of a graph. The formal mathematical definition can be found in calculus texts.

Exercise 9.2

The graph below shows the population of a certain region in the course of a century. Find the derivative at $t = 25$ years. What does the derivative tell you? Answer without using the terms slope, rate, or derivative.

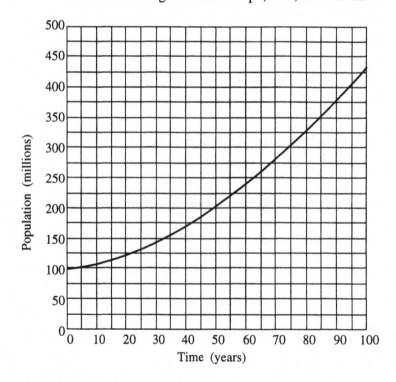

Exercise 9.3

A hiker climbs a mountain that has a peak altitude of 8000 ft above sea level. As the hiker gradually ascends to the top, she measures the air temperature at many different altitudes. Her measurements are shown in the graph below.

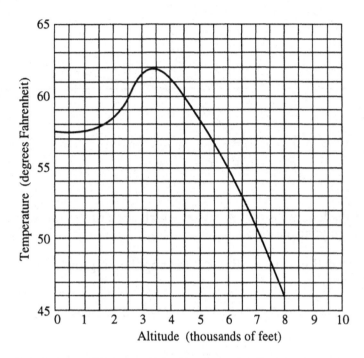

Find the derivative of the graph at the point that corresponds to altitude 5500 ft. What does this derivative tell you? Answer without using the terms slope, rate, or derivative.

✔ Check your reasoning with a staff member.

Sometimes information about the behavior of a quantity must be obtained from a graph of its rate of change. How this can be done quantitatively is discussed in Sections 12 and 13. At this stage, however, we can determine several qualitative features of the behavior of a quantity from a graph of its rate of change.

McDermott & P.E.G., U.Wash./ *Physics by Inquiry*

Exercise 9.4

Below is a graph of population growth rate versus time for the fictitious country of Lauternstein.

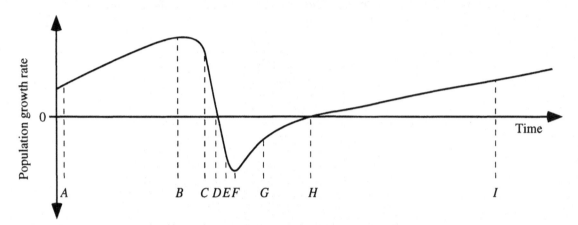

List all of the lettered times, if any, at which each of the following is true. Explain your reasoning.

A. The population is increasing but the population growth rate is decreasing.

B. The population is decreasing and the rate of decrease is becoming larger.

C. The population is increasing and the rate of growth is negative.

D. The population is increasing and the rate of growth is also increasing.

E. The population is decreasing but the rate of decrease is diminishing.

F. The population is neither increasing nor decreasing.

G. The population is changing most quickly.

H. The population is maximum.

I. The population is minimum.

J. The population growth rate is neither increasing nor decreasing.

✔ Check your reasoning with a staff member.

Section 10. The concept of acceleration

It is essential to remember that acceleration tells only about change in velocity and gives absolutely no information about the velocity itself. Whenever the acceleration is in question, a good first step is to focus on changes in velocity. An equally important step is to consider how much time passes during the change in velocity, because acceleration is the rate of change of velocity. In symbols it is:

$$a = \frac{\Delta v}{\Delta t} \qquad \text{(if } a \text{ is constant)}$$

Experiment 10.1

Set up two four-foot sections of track so that the first section is inclined and the last section is level.

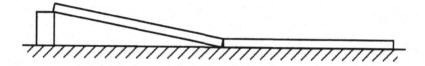

Start three clocks running synchronously. Release a ball from rest at the top of the inclined section and let it roll. Stop one clock when the ball is released, another when the ball reaches the beginning of the level section, and the third when the ball reaches the end of the level section. Compute the acceleration on the inclined section of track.

✔ Check your reasoning with a staff member.

Experiment 10.2

Have a staff member show you the demonstration for this experiment.

Exercise 10.3

Tell whether each of the following statements is *always* true. If the statement is not always true, give an example in which it is not true.

A. Two objects are traveling in the same direction on parallel tracks. They have different speeds, and object A is ahead of object B. Object A is speeding up and maintains a constant acceleration. If object B maintains a constant speed, it will never overtake (pass) object A.

B. If two objects have constant acceleration over the same time interval, the one with the greater increase in velocity must have the greater acceleration.

C. If two objects moving with constant acceleration cover the same distance in the same amount of time, they must have the same acceleration.

D. If two objects have constant acceleration and they change velocity by the same amount, the one that does it in less time must have the larger acceleration.

Exercise 10.4

Below is a position versus time graph for two objects that move with constant acceleration. Which object has the greater acceleration? Explain your reasoning.

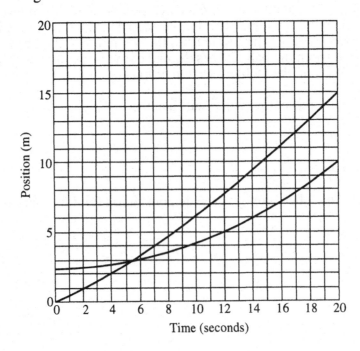

✔ Check your reasoning with a staff member.

McDermott & P.E.G., U.Wash./*Physics by Inquiry*

Free fall

One of the most frequently encountered accelerated motions is *free fall*. A ball that is dropped or thrown is said to be in free fall. The following two experiments illustrate two properties of free fall motion.

Experiment 10.5

A. Mount a ticker timer so that the tape can pass through it vertically. Pass about 60 to 70 cm of tape through the timer and attach a mass (100 to 300 g) to the bottom of the tape. Hold the top end of the tape so that the tape hangs through the timer. The bottom of the tape should be in position to be struck by the timer.

Turn on the timer and release the tape. Plotting points for every 1/60 second, plot each of the following graphs:

(1) position versus time

(2) velocity versus time

(3) velocity versus position

B. There are at least two ways we might describe how quickly an object speeds up as it falls. We could calculate (1) the increase in velocity in each second elapsed, or (2) the increase in velocity in each centimeter fallen.

Based on your experiment, which description do you think provides more insight into the motion?

Experiment 10.6

Obtain several objects and drop them simultaneously from the same height. What effect does mass have on the way objects fall?

✔ Check your results with a staff member.

Sign of the acceleration

The sign of the velocity has a simple interpretation. The velocity gets its sign from the Δx in $\Delta x/\Delta t$. A positive velocity means Δx is positive and therefore that the object is moving in the positive direction.

The sign of the acceleration is not as easy to interpret. As we have seen in Exercise 7.3, a negative acceleration can indicate either speeding up or slowing down. To find out whether the object is speeding up or slowing down, we must know the sign of the velocity and the sign of the acceleration.

Exercise 10.7

For each graph indicate:

- the sign of the velocity,

- whether the velocity is increasing or decreasing algebraically,

- the sign of the acceleration, and

- whether the object is speeding up or slowing down.

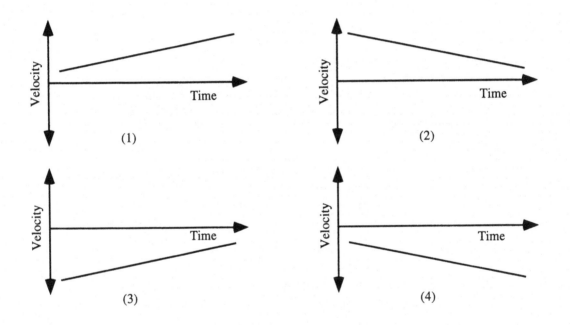

As can be seen from the exercise above, the object is speeding up if the velocity and acceleration have the same sign, whether positive or negative. If the velocity and acceleration have different signs, then the object is slowing down.

Negative acceleration does not necessarily mean slowing down.

Positive acceleration does not necessarily mean speeding up.

Exercise 10.8

If the acceleration is negative, then the change in velocity, Δv, is negative because the acceleration gets its sign from the Δv in $\Delta v/\Delta t$. If the velocity is also negative, then the velocity is becoming a larger negative number, and then the object is speeding up. To which of the four graphs in Exercise 10.7 does this description apply?

Exercise 10.9

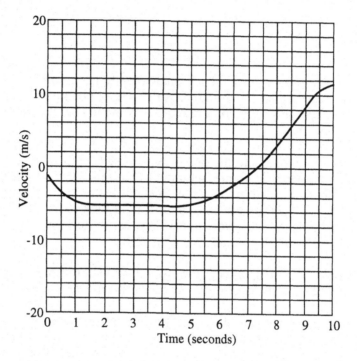

Answer the following questions based on information given in the graph above. Tell how you decided on your answer.

A. At what times, if any, did the object have positive acceleration and negative velocity at the same time?

B. At what times, if any, did the object have negative acceleration and positive velocity at the same time?

C. At what times, if any, was the acceleration zero?

D. At what times, if any, was the object speeding up?

E. At what times, if any, was the object slowing down?

F. At what times, if any, did the object sit still for an extended period of time?

✔ Check your results with a staff member.

Analogies with acceleration

Sample problem

When a city grows, it uses more water. A particular city gets all of its water from a federal reservoir. The reservoir maintains a water meter that counts the number of gallons that pass into the city. From meter readings, it has been determined that the rate of water use is increasing by 50,000 gal/day each year.

Make a mathematical analogy between the quantities related to the water use in this city and quantities from motion.

Sample solution

The quantity 50,000 gal/day each year is analogous to acceleration. One can guess that this is so, because the number 50,000 is the rate of change of a rate of use, and acceleration is the rate of change of a rate of movement. Let us see how the two sets of quantities correspond.

(1) Velocity corresponds to the rate of water use (i.e., how many gallons are used per day). Both are obtained by dividing: Velocity is displacement divided by duration. Rate of water use is amount of water used divided by the amount of time it took to use it. Both must be computed over short time intervals because both ratios vary with time. In short, both are instantaneous ratios.

(2) Displacement corresponds to the amount of water used. Both are in the numerator in the divisions referred to in part 1.

(3) Position is analogous to the meter reading at the reservoir. Positions are subtracted to give displacement; meter readings are subtracted to give amount of water used. Since displacement and amount of water used are analogous, position and meter reading must also correspond to one another.

(4) Duration in the one case corresponds to duration in the other. Duration is in the denominator when both velocity and water use rate are computed, and also in the calculation of acceleration and the rate of change of the water use rate.

(5) Time in the one case corresponds to time in the other. Time is related to duration in the same way in both systems: duration equals final time minus initial time.

(6) Acceleration is analogous to the rate of change of the water use rate, which is 50,000 gal/day per year in this problem. Since velocity and the water use rate are analogous, their rates of change must also be analogous. Their rates of change are acceleration and the rate of change of the water use rate (50,000 gal/day/yr), respectively.

Exercise 10.10

In many countries, the amount of land under cultivation has been decreasing. In a certain country the situation has been very severe due to both urban expansion into farm land and growth of the desert. The rate of loss of agricultural land was as high as 50,000 km^2/yr several years ago. Fortunately, government action has led to a steady decrease in the land loss rate of 100,000 km^2/yr/yr today.

The quantity 100,000 km^2/yr/yr can be considered to be analogous to acceleration. Make this analogy by telling what corresponds to each of the following and telling how they are alike.

A. velocity

D. duration

B. displacement

E. time

C. position

F. acceleration

✔ Check your reasoning with a staff member.

Part D: Algebraic representations of motion

In Part D of *Kinematics,* we examine how to represent motion algebraically and how to interpret the kinematical equations. We introduce the process of integration as a means for interpreting the area under motion graphs. The concept of average velocity is developed and applied to uniformly accelerated motion.

Section 11. Interpreting algebraic equations

Before beginning quantitative reasoning with nonuniform motion, let us review the way in which algebra is applied to physics. Before an algebraic equation can be written, there must be an interpretation of the expressions on each side of the equation and a justification of why they are set equal. This procedure is not followed in the example below, and the result is an incorrect answer. Note that the error made seems reasonable until examined closely.

Sample problem

A uniformly accelerated object has an initial position of 10 cm. The object is at position 24 cm after 2 seconds. How much longer will it take the object to travel 70 cm farther?

Sample solution

Two intervals are involved:

$$\frac{\Delta x_1}{\Delta t_1} = \frac{\Delta x_2}{\Delta t_2}$$

$$\frac{14}{2} = \frac{70}{\Delta t_2}$$

$$\boxed{\Delta t_2 = 10 \text{ seconds}}$$

Comments

The very first line of the solution is incorrect. Although the two sides of the equation look very similar, there is no reason to believe these two numbers are the same. In a proper solution, there should be an interpretation of each side of the equation. There is no simple interpretation of the numbers $\Delta x/\Delta t$ in nonuniform motion; the distance is not shared equally by the units of time. There is no justification for setting the number $\Delta x_1/\Delta t_1$ equal to the number $\Delta x_2/\Delta t_2$ —indeed they are not equal. The problem as stated cannot be solved at all; there is not enough information provided.

Exercise 11.1

In each of the following, a problem and a solution are given. Tell whether the solution is correct or incorrect. If it is correct, tell why. If it is incorrect, tell why.

A. *Problem:* An object moving with uniform acceleration increases speed by 60 cm/s in 2.0 seconds. How much longer will it take to increase speed by another 100 cm/s?

Proposed solution:

$$\frac{\Delta v_1}{\Delta t_1} = \frac{\Delta v_2}{\Delta t_2}$$

$$\frac{60}{2} = \frac{100}{\Delta t_2}$$

$$\boxed{\Delta t_2 = 3.3 \text{ seconds}}$$

B. *Problem:* An object speeding up with constant acceleration has initial position 120 m and initial velocity 3.0 m/s at time $t = 180$ seconds. Where will the object be at $t = 200$ seconds?

Proposed solution:

$$\Delta x = v\Delta t$$

$$x_2 - x_1 = v(t_2 - t_1)$$

$$x_2 - 120 = 3(200 - 180)$$

$$\boxed{x_2 = 180 \text{ m}}$$

C. *Problem:* A square plot of land with an area of 12 acres is 220 m across. How far across is a square plot of land with area 18 acres?

Proposed solution:

$$\frac{A_1}{l_1} = \frac{A_2}{l_2}$$

$$\frac{12}{220} = \frac{18}{l_2}$$

$$\boxed{l_2 = 330 \text{ m}}$$

✔ Check your reasoning with a staff member.

The relation between position and time is more complicated in nonuniform motion than it is for uniform motion. Since $\Delta x/\Delta t$ is not the same for various parts of the motion, Δx and Δt are not proportional. The motion must be considered bit by bit in small intervals. There is no simple algebraic formula for finding the position of an object at an arbitrary time that will work for all nonuniform motions. We cannot generally write an algebraic expression for x in terms of t.

Section 12. Determining displacement for uniformly accelerated motion

There is a general method of finding the relation between position and time for any nonuniform motion, but it is not algebraic. The method is nicely represented on a graph. We will work entirely in the context of velocity versus time graphs.

Suppose we have the v versus t graph for an arbitrary nonuniform motion; for example, the one below.

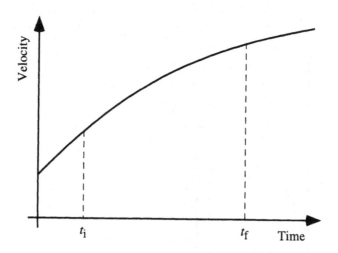

We know that within small intervals the motion is very nearly uniform. We will therefore break up the motion into many very small segments and replace each one with a short section of uniform motion. The actual motion differs very little from this series of uniform motions that replace the preceding graph.

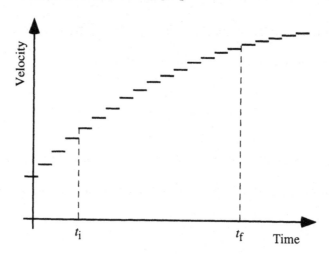

We will call the duration of the small intervals δt and the displacement in the small intervals δx. (δ is the lower-case Greek letter *delta*. It is often used to represent very small changes.) In each small interval, the motion is uniform, so the displacement in each of these intervals of uniform motion will just be the velocity in the interval multiplied by the duration: $\delta x = v\delta t$ for each small interval. The velocity is represented on the graph by the height of the graph above the t-axis. The duration, δt, is represented by the width of the interval along the t-axis. These two lengths are shown on the graph following.

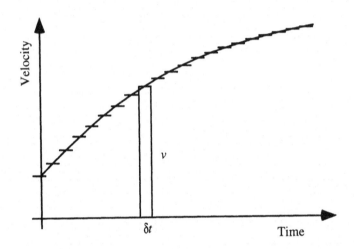

When we calculate the displacement, δx, of the object in this small interval, we multiply the velocity times the duration. This corresponds to multiplying the height and width of the tall, thin rectangle in the preceding graphs. Therefore, the area of this small rectangle represents the displacement during the interval:

$$\delta x = \text{the area of the rectangle.}$$

The height of the rectangle represents a velocity, not a distance, and the width represents a duration; so the area represents a displacement, not a number of square centimeters.

McDermott & P.E.G., U.Wash./ *Physics by Inquiry*

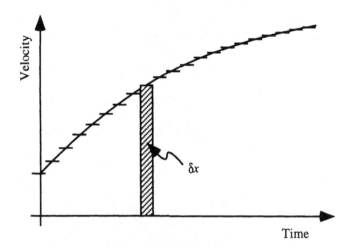

In order to figure out the entire displacement over the whole motion in question, we just add up all of the δx's from each of the small intervals:

$$\text{Total displacement} = \delta x_1 + \delta x_2 + \delta x_3 + \delta x_4 + \ldots \text{ to the end.}$$

Graphically, this sum is represented by adding the areas of the tall, thin rectangles, as indicated below.

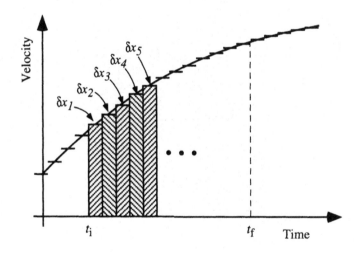

The total displacement for the entire motion in question is represented by the total area of all the rectangles. Each rectangle contributes the displacement during one of the small intervals of motion. The following graph shows the areas of all of the thin rectangles combined. This total area represents the total displacement from t_i to t_f.

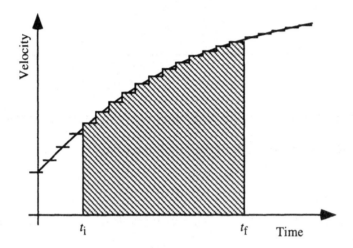

If we divide the motion into even smaller intervals, the small segments of uniform motion will represent the actual motion even more faithfully. The steps along the graph become very fine, almost as smooth as the graph itself. The following graph shows the area under the graph when the rectangles are so small that they cannot be distinguished.

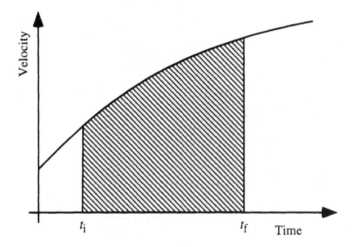

The point is that it is not necessary to actually draw small rectangles when finding the area under the graph. In fact, it is better to just find the area under the original graph itself. Using the graph itself is just like using very many tiny rectangles rather than a few large ones, and it is therefore more accurate. We have now arrived at a major conclusion:

The area under the graph of velocity versus time
for an interval of motion gives the displacement during that interval.

If the velocity of a moving object is negative, its displacement will be negative also. We must take care that the area between the graph and the *t*-axis is treated as a negative number when the velocity is negative. A good rule to remember is:

Areas above the t-axis are positive;

areas below the t-axis are negative.

Sample problem

Find the displacement from *t* = 2 seconds to *t* = 8 seconds for the motion represented on the following graph.

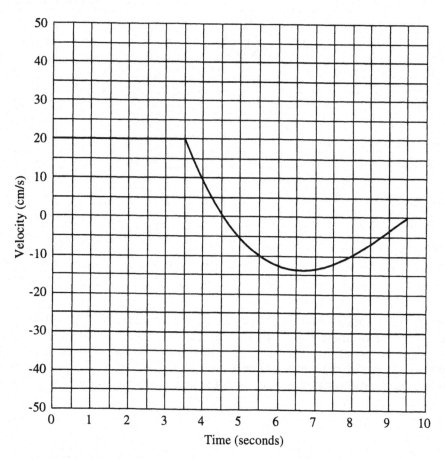

Sample solution

First, we must decide how to get information about displacement from the graph. The height will not tell us—height represents velocity. The slope will not work either; the slope would tell us $\Delta v/\Delta t$, which is the acceleration. What about the area under the graph? To see whether the area can tell us anything, we must think about a tall, thin rectangle under the graph: The height of such a rectangle represents velocity, and the width represents the duration of a short interval. The area of the rectangle (height times width) would be v times δt, which is indeed the displacement in that interval.

The heights of the rectangles must represent velocities, so the rectangles should extend from $v = 0$ to the graph of the motion. The rectangles should *not* start at the bottom of the graph paper. We really want the area between the graph and the t-axis. The problem asks for the displacement from $t = 2$ seconds to $t = 8$ seconds, so we should find the area between the graph and the t-axis beginning at $t = 2$ seconds and ending at $t = 8$ seconds.

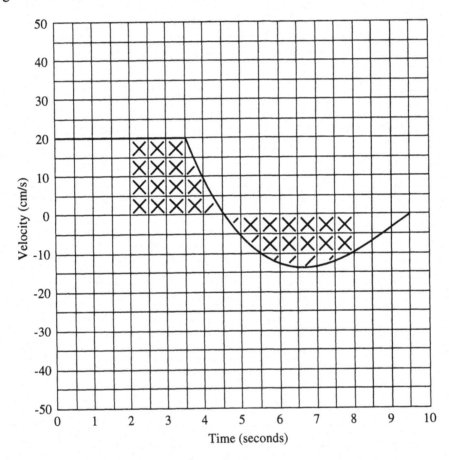

To find the area under the graph, we count the squares and partial squares. As shown on the graph above, there are about 16 squares above the t-axis from $t = 2$ seconds to $t = 4.5$ seconds. After that, the velocity is negative. The squares below the t-axis represent negative displacement. There are approximately 14.5 squares between the graph and the t-axis from $t = 4.5$ seconds to $t = 8$ seconds.

Thus, the total displacement is represented on the graph by 16 – 14.5 = 1.5 squares. But how many centimeters do 1.5 squares represent? This depends on the scales used on both axes of the graph. We can find how much displacement one square represents by looking at the height and width of one square.

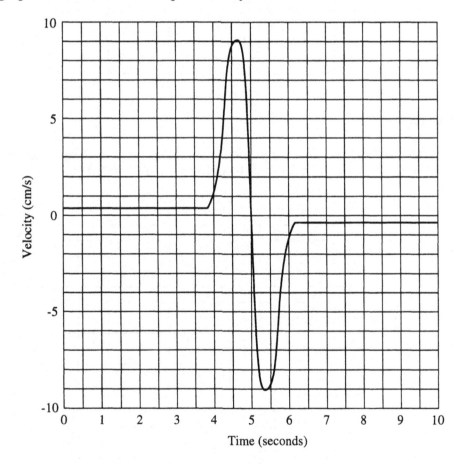

In this problem, the area of one square is given by 5 cm/s × 0.5 seconds, or 2.5 cm. The displacement is thus 2.5 cm/square × 1.5 squares, which is 3.8 cm.

Exercise 12.1

Below is the graph of velocity versus time for a moving object. Assume the object is at $x = 0$ cm at $t = 0$ second and plot a position versus time graph for this motion. Plot points every 0.5 second.

✔ Check your graph with a staff member.

Exercise 12.2

Below is a velocity versus time graph for an object that is located at $x = 0$ when $t = 0$. At what times is the object located at $x = 110$ cm?

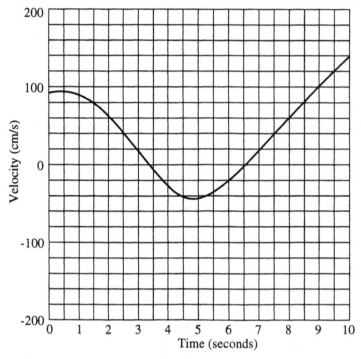

Exercise 12.3

Below is a velocity versus time graph for an object. At $t = 10$ seconds, the object is at $x = 500$ cm. When did the object pass by $x = 250$ cm?

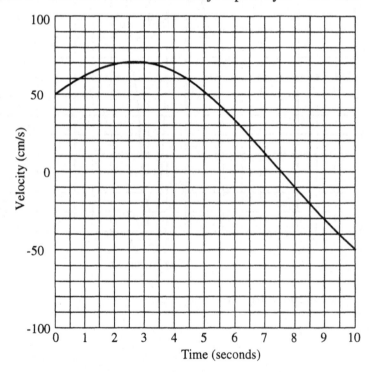

McDermott & P.E.G., U.Wash./ *Physics by Inquiry*

Exercise 12.4

Below is the velocity versus time graph for an object that moved from
$x = 0$ cm at $t = 0$ seconds to $x = 1000$ cm in 6 seconds. The velocity axis
has not been labeled. If the center of the graph represents $v = 0$, what does
the top line represent?

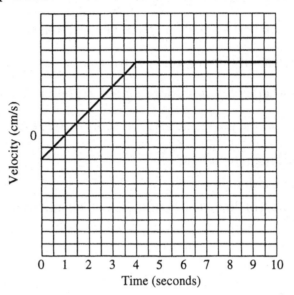

✔ Check your reasoning with a staff member.

Exercise 12.5

At right is a velocity versus time
graph for a car moving from *A* to
B. The car leaves *A* at $t = 0$. *B* is
located 210 m away from *A*.

A stunt man is planning to jump
from a helicopter and land in the
car. For the stunt to work, the car
must be at *B* 10 seconds after
leaving *A* and have a velocity of
20 m/s.

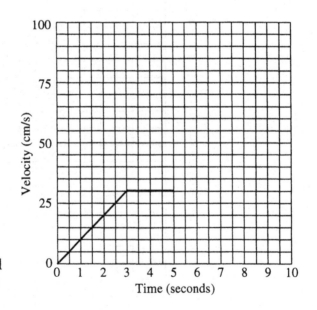

The first half of motion of the car
is shown on the graph. On the
same axes, draw a graph for the
second half of the motion that will
get the car to the right place at the
right time with the right speed.

✔ Discuss your answers with
a staff member.

Section 13. Finding the area under a graph

The process of finding the area under a graph is called *integration*. For a *v* versus *t* graph, the resulting number is called the *integral of the velocity over time*. There are many applications of integration besides finding displacements. Integration can give information about any quantity from a graph of its rate.

In order to decide whether the area under a graph represents anything useful, imagine a tall, thin rectangle under the graph and ask yourself the following questions:

(1) What does the height represent?

(2) What does the width represent?

(3) Does the product (multiplication) of these two quantities have a useful interpretation?

Sample problem

Below is a graph of population growth rate versus time for the republic Forwichistan. If the population of Forwichistan was 28 million at *t* = 15 years what was the population at *t* = 80 years?

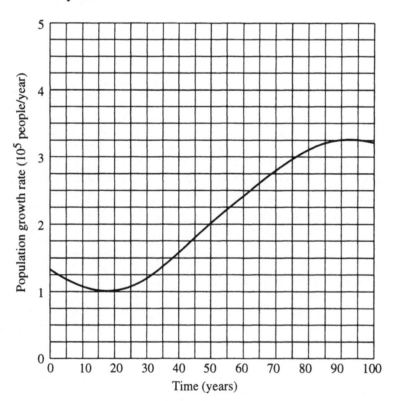

Sample solution

Since we are asked about population but are given a graph of its rate, we will almost certainly need to integrate. To check, however, we draw a tall thin rectangle under the graph and see whether the area of this rectangle represents useful information.

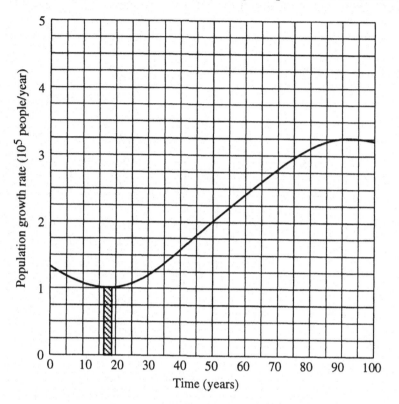

The height of such a rectangle represents the population growth rate, or the increase in population per unit time. The width of a tall thin rectangle represents a small amount of time. The area of the rectangle (height times width) therefore represents the increase in population that occurred during the small interval of time represented by the width of the rectangle.

The increase in population is just what is required to solve the problem. All we need to do is take the initial population of 28 million and add all of the increases in population that occurred between $t = 15$ and $t = 80$. To determine the total increase in population, we need to find the area under the graph between $t = 15$ and $t = 80$.

A count of the squares gives (by one estimate) 50.8. Each small square on the graph is 0.5×10^5 people/year high and 5 years wide. The area of one square represents an increase of (5 years) $\times$ (0.5×10^5 people/year) $= 2.5 \times 10^5$ people in the population. The increase in population from $t = 15$ years to $t = 80$ years is thus (50.8 squares) $\times$ (2.5×10^5 people/square) $= 12.7$ million people.

To get the population of Forwichistan at $t = 80$ years, we must add the 12.7 million obtained from integrating the graph to the initial population of 28 million, which is not represented on the graph. The answer is thus 40.7 million.

Exercise 13.1

Below is a graph of the acceleration versus time for a ball rolling down an inclined plane.

A. What does the height of a thin rectangle under the graph represent?

B. What does the width of a thin rectangle under the graph represent?

C. Give an interpretation of the area under this graph.

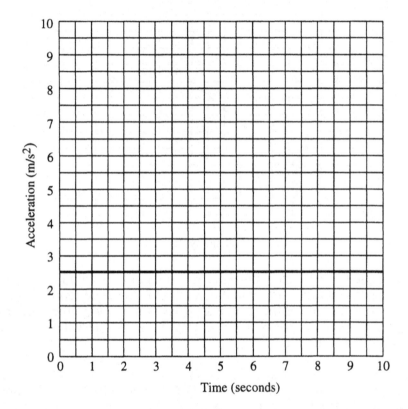

Exercise 13.2

Below is an acceleration versus time graph for a child on an amusement park ride. The child starts from rest.

A. Plot a velocity versus time graph for the entire ride. Plot points every 2.5 seconds.

B. How far did the child travel in the first 10 minutes?

C. Did the child ever go backwards? If so, at what time did the child reverse direction?

D. Is the ride over at $t = 50$ seconds? How can you tell?

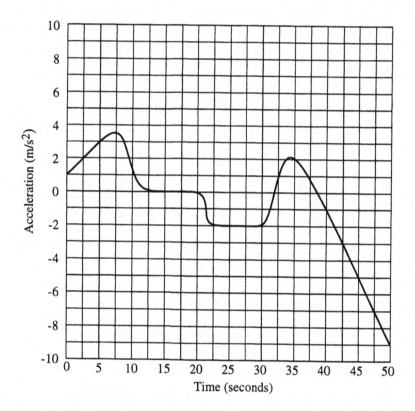

✔ Discuss this exercise with a staff member.

Exercise 13.3

When people analyze the transportation system of a country, they often measure the shipping capacity by telling the total weight of all the ships in that country's service. This measure is sometimes called the "shipping tonnage." Below is a graph of the rate of change of the shipping tonnage of the country of Lauternstein over a hundred-year period. In 1830, the total shipping tonnage of Lauternstein's fleet was 220 million tons.

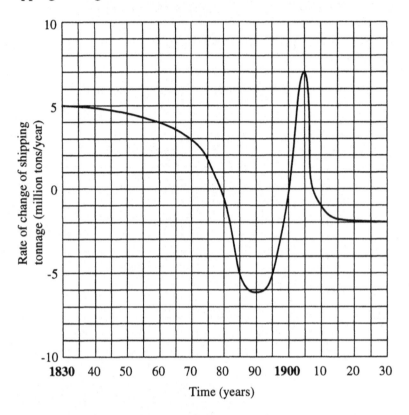

Explain your reasoning on the following questions:

A. When (what year) was the shipping tonnage largest?

B. When was the shipping tonnage smallest?

C. When was the shipping tonnage increasing the fastest?

D. When was the shipping tonnage decreasing the fastest

E. What was the largest rate of increase in shipping tonnage

F. What was the largest rate of decrease in shipping tonnage?

G. What was the largest shipping tonnage during the time shown?

H. What was the smallest shipping tonnage during the time shown?

Exercise 13.4

In some areas of the world, the land is slowly subsiding or sinking. Often
this results from underground water being taken up to the surface through
wells. Below is a graph that shows the variation of the rate of subsidence
of the land in a hypothetical region over a century.

In what year was the level of the land one meter below its 1890 level?

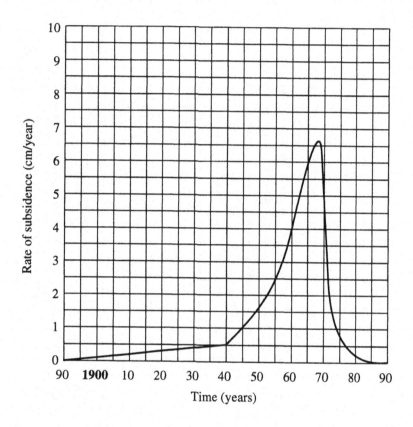

✔ Discuss this exercise with a staff member.

McDermott & P.E.G., U.Wash./*Physics by Inquiry*

Exercise 13.5

A number of rivers flow into a bay that has only a small opening to the sea. The water in this bay is salty on the bottom and nearly fresh at the surface. The graph below shows the variation of salt concentration with depth.

An open pipe with cross-sectional area 1000 cm^2 is lowered vertically into the bay and fills with bay water as it is lowered. If the top 1×10^6 cm^3 of water in the pipe are pumped into a tank, how much salt will be in the tank?

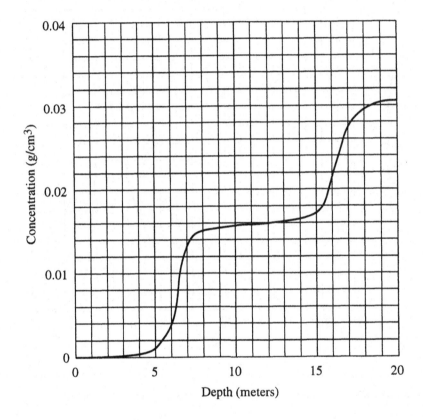

McDermott & P.E.G., U.Wash./ *Physics by Inquiry*

Exercise 13.6

The graph below shows the variation of the rate of change of the national debt of a hypothetical country. Positive values mean the debt is increasing, negative values mean the debt is decreasing. In 1830 the national debt was $\$4 \times 10^8$. In 1900 the national debt was half of what it had been in 1830.

A. Use the above information to determine the scale for the vertical axis.

B. What was the maximum rate of change of the national debt?

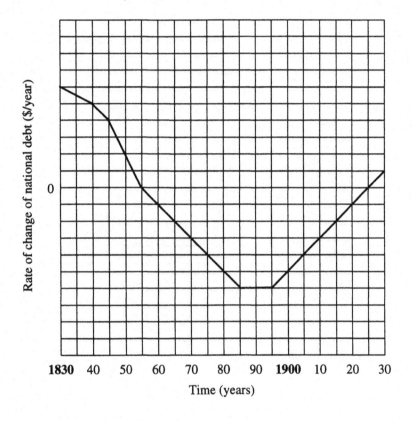

Exercise 13.7

The graph below shows the elevation gain per 100 meters for a road in the mountains plotted against the distance from a town in the valley. From the town, the road ascends into the mountains and then comes down to a flat-bottomed valley.

A. How far from the town is the pass?

B. If the elevation of the flat-bottomed valley is 500 meters, what is the elevation of the pass?

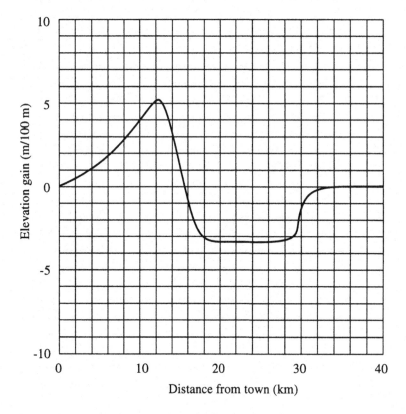

✔ Discuss this exercise with a staff member.

Section 14. Calculating averages

Averages of continuously varying quantities can frequently give insights into complicated matters without requiring detailed information. Suppose, for example, we want to know something about the climate in a particular place. It would be helpful to know the average temperature during the winter and summer, but it is really not necessary to know the temperature every hour of the day throughout a whole year.

In this section we will deal with the problem of finding average values for quantities like temperature or velocity that vary continuously. Before doing so, however, we will first review the idea of weighted average.

Sample problem

A student received the following grades one quarter: 4.0 in a 5-credit course, 3.0 in a 4-credit course, 3.0 in a 1-credit course, and 2.0 in a 3-credit course. What was the average grade received by the student?

Sample solution

It would be improper to just average the grades in these four courses by adding the grades and dividing by four like this: (4 + 3 + 3 + 2)/4 = 3.0. This procedure is not adequate because it does not reflect the fact that the 5-credit course *carries more weight* than the 3-credit course.

The kind of average most often used for grades is called the grade-point average. It is computed as follows: multiply each grade by its weight in credits, add, and then divide by the total number of credits:

$$\text{gpa} = \frac{(4.0 \times 5) + (3.0 \times 4) + (3.0 \times 1) + (2.0 \times 3)}{5 + 4 + 1 + 3} = \frac{41}{13} = 3.15$$

An average computed in this way is called a *weighted average.*

Exercise 14.1

In a certain house live people of the following ages: six 2-year-olds, four 5-year-olds, one 23-year-old, and one 28-year-old. What is the average age of these people?

Average velocity

It is sometimes desirable to know the average velocity of an object moving with nonuniform motion. For example, we may wish to make an approximate calculation without the detailed analysis necessary to analyze a motion exactly. As we will see later, estimating the time required for an object to move a certain distance is easy if one has an average value for the velocity, even though the velocity may be changing. Although the usefulness of an average lies in simplifying problems, the definition of an average of a continuously varying quantity like velocity is not simple and requires careful thought.

To find the average velocity of an object, we must take into account all of the different velocities the object has and also how much time it has each of these velocities. Consider the case of two cars that travel for one hour. Suppose that both cars begin the trip with a speed of 20 mph and end with a speed of 50 mph, but that the second car speeds up earlier than the first. If the first car spends 55 minutes going 20 mph and 5 minutes going 50 mph, it will spend most of the time traveling slowly and will have a small average velocity. If the second car spends 5 minutes going 20 mph and 55 minutes going 50 mph, it will have a large average velocity. Thus both the velocities the car has and also how long it has each velocity are important in determining the average velocity.

We take both of these elements into account in a procedure that is analogous to what we do when taking a weighted average. We average all of the many velocities the moving object has, but we weight each velocity by the amount of time it has this velocity. The calculation we perform is the following:

$$v_{av} = \frac{v_1 \delta t_1 + v_2 \delta t_2 + v_3 \delta t_3 + v_4 \delta t_4 + \ldots}{\delta t_1 + \delta t_2 + \delta t_3 + \delta t_4 + \ldots}$$

The method of computation appears straightforward. However, care must be taken because there may be many velocities, and the times that we should use to weight the velocities may be very short. This situation is tailor-made for integration, however, and we can average the velocity easily by applying what we know about integration.

Suppose, for example, we want to find the average velocity over the interval shown in the graph below. The velocities are represented by heights on the graph and the times are represented by small intervals along the *t*-axis. The products $v_1\delta t_1$, $v_2\delta t_2$, and so on, are therefore represented by the areas of thin rectangles under the graph. The sum of these terms is thus the integral of the velocity over time. In the formula for average velocity:

$$v_{av} = \frac{v_1\delta t_1 + v_2\delta t_2 + v_3\delta t_3 + v_4\delta t_4 + \ldots}{\delta t_1 + \delta t_2 + \delta t_3 + \delta t_4 + \ldots}$$

The numerator thus is the integral of velocity over time. The denominator is just the sum of all the little time intervals, which is the total time:

$$v_{av} = \frac{\text{the integral of the velocity over time}}{\text{the time of motion}}$$

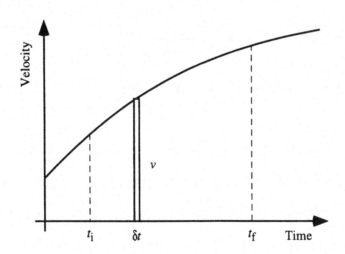

Exercise 14.2

A. Find the average velocity from $t = 0$ to $t = 10$ seconds for the motion on the graph below.

B. On the same axes, plot a graph that corresponds to uniform motion with the average velocity.

C. How do the areas under the two graphs compare for the interval from $t = 0$ to $t = 10$ seconds?

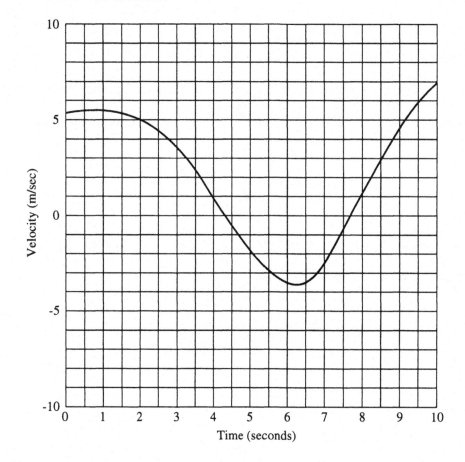

✔ Check your reasoning with a staff member.

As the exercise above illustrates, the averaging process can be thought of as a leveling-out of a graph so that the area underneath is still the same, just as earth may be moved from hills to valleys to produce a level roadway. Thinking of averaging as the leveling-out of a graph can provide a way of estimating averages "by eye." This can be done by holding a straight-edge horizontally across the graph and sliding the line up and down until the area under the line and under the graph appear to be the same.

Exercise 14.3

Estimate the average velocity for each of the motions on the graphs below over the entire intervals shown. Draw a horizontal line representing your estimate of the average velocity.

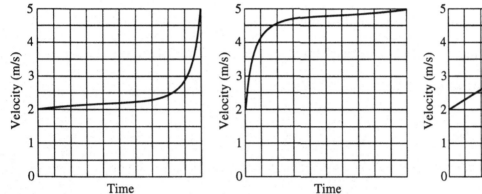
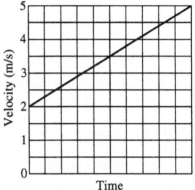

The exercise above illustrates how important weighting is in averaging velocity. Each motion has the same range of velocities, beginning at $v = 2$ m/s and ending at $v = 5$ m/s. The motions differ in how long the objects move at each speed. One object spends a long time at low speeds and thus has a small average velocity. Another object spends a long time at high speeds and has a large average velocity. The third object speeds up uniformly, favoring neither high nor low speeds and has an intermediate average velocity.

Consider the procedure involved in finding the average of velocity over time. We must calculate the numerator and denominator in the formula:

$$v_{av} = \frac{v_1 \delta t_1 + v_2 \delta t_2 + v_3 \delta t_3 + v_4 \delta t_4 + \ldots}{\delta t_1 + \delta t_2 + \delta t_3 + \delta t_4 + \ldots}$$

As we have discussed, the numerator is the integral of velocity over time. This quantity is *also* the total displacement, Δx, of the moving object during the interval in question. The denominator is the total duration, Δt, of the interval. The average velocity can thus be obtained easily by dividing total displacement by total time:

$$v_{av} = \frac{\Delta x}{\Delta t}$$

This is an important result and is very useful in solving certain kinds of problems.

          McDermott & P.E.G., U.Wash./ *Physics by Inquiry*

It has been stressed several times, most recently at the beginning of *Kinematics,* that the number $\Delta x/\Delta t$ has no simple interpretation if the motion is nonuniform. As we see now, however, it can be interpreted as the average velocity of the moving object. While the intuitive concept of average is not difficult, the meaning of an average of a continuously varying quantity like velocity involves integration and therefore should perhaps not be considered simple.

Uses of average velocity

A second interpretation of average velocity is that $\Delta x/\Delta t$ is the velocity the object *would* have *if* it had the same displacement in the same amount of time but moved with uniform motion. The average velocity is especially useful in doing approximate calculations with motions that are repetitive. Both of these ideas are illustrated in the exercise below.

Exercise 14.4

On an automobile trip, a traveler drives at a speed of 55 mph for two hours and then stops for half an hour to rest. He then repeats this behavior over and over again.

A. What is the traveler's average velocity?

B. How long will it take to complete a 600-mile journey? (Just give an estimate.)

C. Make a graph of instantaneous velocity versus time for this motion. On the same set of axes, plot a graph that corresponds to uniform motion with the average velocity. How do the areas under these two graphs compare?

D. Make a position versus time graph for this motion. On the same set of axes, plot a graph that corresponds to uniform motion with the average velocity.

✔ Check your reasoning with a staff member.

Averages of other continuously varying quantities

The techniques used to find the average velocity of a moving object can be applied to

find the average of other continuously varying quantities. Examples are provided below.

Sample problem

A recording thermometer that continuously charts the outdoor temperature gave the data plotted below. Find the average temperature for the 24-hour period shown.

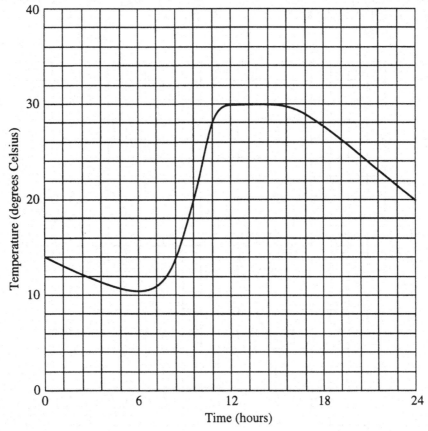

Sample solution

If the temperature is low for most of the day, the average temperature should be low also. In other words, the temperatures must be weighted by the amount of time each temperature was held. To compute the time-weighted average of the temperature, we must compute:

$$T_{av} = \frac{T_1 \delta t_1 + T_2 \delta t_2 + T_3 \delta t_3 + T_4 \delta t_4 + \ldots}{\delta t_1 + \delta t_2 + \delta t_3 + \delta t_4 + \ldots}$$

As before, the sum in the numerator can be interpreted as the integral of temperature over time. Each quantity $T_1 \delta t_1$, $T_2 \delta t_2$, and so forth, represents the area of a tall, thin rectangle under the temperature versus time graph. The total sum thus represents the integral, and the denominator is just the total time, 24 hours.

McDermott & P.E.G., U.Wash./*Physics by Inquiry*

The area under the graph is about 208 squares. Each square has a height of 2°C and a width of 1.2 hours and thus its area represents 2.4°C•h. The integral of temperature over time is therefore 208 squares × 2.4°C•h/square = 499°C hours. The average temperature is given by this integral divided by the elapsed time of 24 hours:

$$T_{av} = 499°C•h/24\ h = 20.8°C$$

If we plot a constant temperature of 20.8°C on the same axes as the graph of the actual temperature, we see again how averaging by integration corresponds to a leveling-out of the peaks and valleys of the graph.

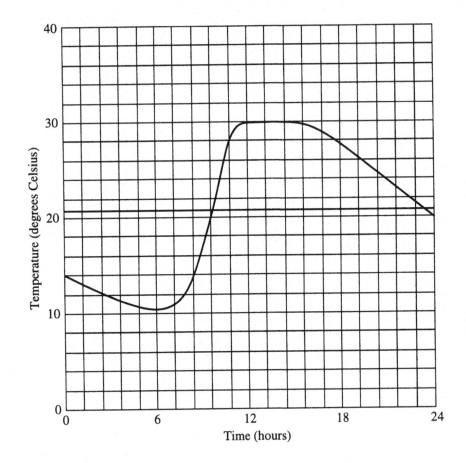

Exercise 14.5

An oceanography student measures the salinity of a certain body of sea water at many different depths. These measurements are represented in the graph below. If someone asks how salty this particular body of water is, we should probably answer with an average value. Find the average salinity of the sea water from the information in the graph below.

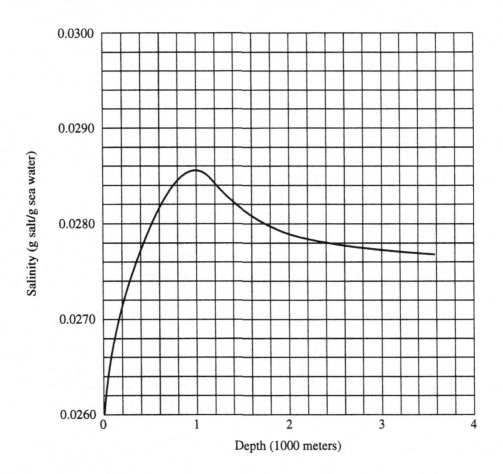

The exercise below illustrates how averaging can make it possible to obtain some information about a complicated system without having to do a detailed analysis. Suppose we wanted to find the average density of an object composed of several substances. An object composed mainly of dense materials should have a high density. Therefore, the average density should be weighted according to the amount of material. We could compute this average as follows:

$$D_{av} = \frac{D_1V_1 + D_2V_2 + D_3V_3 + D_4V_4 + \dots}{V_1 + V_2 + V_3 + V_4 + \dots}$$

The expression in the numerator is the integral of the density over the volume. But each term in the sum in the numerator, D_1V_1, D_2V_2, and so on, is just the mass of material contained in the small volume V_1, V_2, and so on. The numerator is thus just the sum of all the mass in the object: the total mass. Since the denominator is the total volume, the formula for average density reduces to:

$$D_{av} = \frac{M_t}{V_t}$$

Just as was the case with averaging velocity over time, averaging density over volume does not require that an integration be performed.

Exercise 14.6

The total mass of the earth is estimated from its effect on the motion of the moon and satellites to be 6.0×10^{27} g. The volume is estimated by surveying to be 1.1×10^{27} cm^3.

A. What is the average density of the earth?

B. Rocks and soil on the surface of the earth generally have densities around 3 g/cm^3. From this fact and the average density of the earth calculated in part A, we can say something about the material in the interior of the earth even though no one has ever seen this material. What can we say?

✔ Check your reasoning with a staff member.

Section 15. Relating graphs to algebraic equations

Sometimes a particular kind of motion can be integrated to yield an algebraic expression for the position in terms of time. This is the case with uniformly accelerated motion (motion with constant acceleration). Consider the following graph of a uniformly accelerated motion:

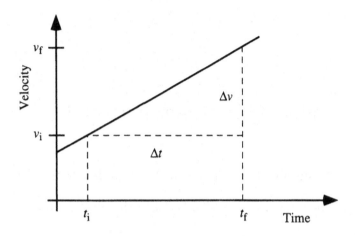

The area under the graph between t_i and t_f is shown above. Notice that the area can be split into two parts, a rectangle and a triangle.

Rectangle: The height of the rectangle is v_i and the width is Δt. Therefore, the area of the rectangle is $v_i \Delta t$.

Triangle: The height of the triangle is Δv, which is also equal to $a \Delta t$. The width of the triangle is Δt. The area of any triangle is one-half of the width times the height, so the area of the triangle is $\frac{1}{2}(a\Delta t)(\Delta t)$ or $\frac{1}{2}a(\Delta t)^2$.

The area under the velocity versus time graph gives the displacement from t_i to t_f. Therefore, we have just derived the following expression for the displacement of a uniformly accelerated object:

$$\Delta x = v_i \Delta t + \frac{1}{2}a(\Delta t)^2 \tag{15.1}$$

Sometimes a different notation is used when $t_i = 0$. In this special case, the subscript "0"

is used to denote initial quantities, and final quantities are written without subscripts.

$$x - x_0 = v_0 t + \frac{1}{2} at^2 \qquad (15.2)$$

Thus x_i becomes x_0, v_i becomes v_0, t_f becomes t, x_f becomes x, and v_f becomes v. We

then have:

$$x = x_0 + v_0 t + \frac{1}{2} at^2 \qquad (15.3)$$

Exercise 15.1

Give an interpretation of each of the terms in equation 15.3:

A. x_0

B. $v_0 t$ (This number is a displacement. What displacement?)

C. $\frac{1}{2} at^2$ (This number is a displacement. What displacement?)

For uniformly accelerated motion, the average velocity is related to the initial and

final velocities in an especially simple way. If we compute the average velocity of the

motion shown on the graph at the beginning of this section, we find that it falls midway

between the initial and final velocities. The following diagram illustrates the process of

finding the average by drawing a level line that has the same area under it as the actual

velocity versus time graph.

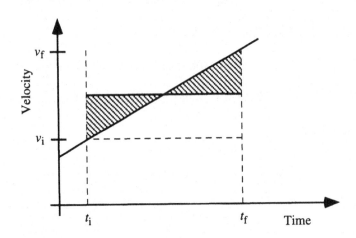

Because the velocity versus time graph is a straight line, the shaded areas will be equal only when the horizontal line representing the average velocity passes through the midpoint of the velocity versus time graph. The average velocity for this uniformly accelerated motion is therefore:

$$v_{av} = \frac{1}{2}(v_i + v_f), \text{ or}$$

$$\frac{\Delta x}{\Delta t} = \frac{v_i + v_f}{2}$$

Exercise 15.2

Use $\Delta x/\Delta t = \frac{1}{2}(v_i + v_f)$ and $v_f - v_i = a\Delta t$ to derive $\Delta x = v_i \Delta t + \frac{1}{2}a(\Delta t)^2$ algebraically.

Hint: Eliminate v_f. If you need help with the algebra, consult your instructor.

Exercise 15.3

Consider an interval during which an object speeds up with uniform acceleration. At the beginning of the interval, the object's instantaneous velocity is less than the average velocity for the whole interval. At the end of the interval, the object's instantaneous velocity is larger than the average velocity. Therefore, the object's instantaneous velocity must be equal to its average velocity sometime in the middle of the interval. Does this happen:

- when $t = \frac{1}{2}(t_i + t_f)$?

- when x $= \frac{1}{2}(x_i + x_f)$?

- both?

- neither?

✔ Check your reasoning with a staff member.

Section 16. Solving problems in kinematics

In this section are several examples of how to solve problems about uniformly accelerated motion. The equations for uniformly accelerated motion we have derived so far are:

$$\Delta x = v_i \Delta t + \frac{1}{2} a (\Delta t)^2 \tag{16.1}$$

$$v_f = v_i + a \Delta t \tag{16.2}$$

$$\frac{\Delta x}{\Delta t} = \frac{1}{2} (v_i + v_f) \tag{16.3}$$

A useful equation for uniformly accelerated motion that does not involve t is:

$$v_f^2 - v_i^2 = 2a\Delta x \tag{16.4}$$

This relation is useful in solving certain problems about uniformly accelerated motion. It does not involve any new information beyond what we already have in Equations 16.1, 16.2, and 16.3. Rather, it can be derived algebraically from these equations.

Exercise 16.1

Derive Equation 16.4 from Equations 16.1 and 16.2. *Hint:* There is no Δt in Equation 16.4, so we have to get rid of Δt in Equation 16.1 and 16.2. The easiest way to do this is to solve for Δt in Equation 16.2 and substitute this expression for the Δt's in Equation 16.1.

✔ Discuss this exercise with a staff member.

Sample problem

A ball is thrown vertically upward from the ground with initial velocity 30 m/s. When will the ball reach a window 40 m above the ground? The acceleration of an object in free fall is 10 m/s².

Sample solution

We can solve this problem using Equation 16.1:

$$\Delta x = v_i \Delta t + \frac{1}{2} a (\Delta t)^2$$

We must first choose which direction we want to be positive. The choice corresponds to setting up a coordinate system (see Section 2). In this case, we let the positive direction be upward.

The choice of coordinate system determines the signs of Δx, v_i, and a. The window is above the ground, so $\Delta x = +40$ m. The initial velocity is upward, so $v_i = +30$ m/s. The acceleration is making the velocity a smaller positive number, so $a = -10$ m/s^2. Putting these values in Equation 16.1, we get:

$$40 \text{ m} = 30 \text{ m/s } \Delta t + \frac{1}{2}(-10 \text{ m/s}^2)(\Delta t)^2$$

We can solve this equation using the quadratic formula. The quadratic formula gives solutions to the following equation, where x is the unknown.

$$ax^2 + bx + c = 0$$

There are two solutions:

$$x = \frac{-b + \sqrt{b^2 - 4ac}}{2a} \quad \text{and} \quad x = \frac{-b - \sqrt{b^2 - 4ac}}{2a}$$

These solutions are usually written together like this:

$$x = \frac{b \pm \sqrt{b^2 - 4ac}}{2a}$$

Before we can use this formula, we must put our equation into the form $ax^2 + bx + c = 0$. We get:

$$-5 \text{ m/s}^2(\Delta t)^2 + 30 \text{ m/s } \Delta t - 40 \text{ m} = 0$$
$$\text{or}$$
$$(\Delta t)^2 - 6 \text{ s } \Delta t + 8 \text{ s}^2 = 0.$$

We thus have $a = 1$, $b = -6$, and $c = 8$. The solutions are:

$$\Delta t = \frac{6 \pm \sqrt{6^2 - (4)(1)(8)}}{2} \text{ seconds}$$

$$t = \frac{6 \pm 2}{2} \text{ seconds}$$

$$\Delta t = 4 \text{ seconds or 2 seconds}$$

Why are there two solutions? The ball will rise and reach the window at $\Delta t = 2$ seconds. However, the ball will continue upward for a while and then fall back down. It will pass the window again at $\Delta t = 4$ seconds on the way down. Thus both solutions tell when the ball reaches the window; the ball passes the window twice.

Suggested format for problem solutions

When a solution is presented in an orderly fashion, it is easy to check for errors. Also, a grader can follow the steps and give partial credit. The following is a suggested format for presenting the solution to the preceding sample problem.

Sample format

$$\text{Given:} \qquad \Delta x = 40 \text{ m}$$

$$v_i = 30 \text{ m/s}$$

$$a = -10 \text{ m/s}^2$$

$$\text{Find:} \qquad \Delta t$$

$$\Delta x = v_i \Delta t + \frac{1}{2} a (\Delta t)^2$$

$$40 \text{ m} = 30 \text{ m/s } \Delta t - 5 \text{ m/s}^2 \, (\Delta t)^2$$

$$(\Delta t)^2 - 6 \text{ s } \Delta t + 8 \text{ s}^2 = 0$$

$$\Delta t = \frac{6 \pm \sqrt{6^2 - (4)(1)(8)}}{2} \quad \text{seconds}$$

$$\text{Answer:} \qquad \boxed{\Delta t = 4 \text{ seconds or 2 seconds}}$$

Checking the units to make sure they come out right

Errors of several kinds are possible when solving uniform acceleration problems. Some kinds of errors upset the natural relations among the units of the quantities involved. It is easy to tell when such a mistake has been made simply by checking the units of each term in the equations used. For example, if the problem asks for the acceleration and the answer comes out like this:

$$a = 2 \text{ m/s,}$$

there has clearly been an error because the units of acceleration must be m/s^2 or something equivalent, not m/s.

Sample problem

An object with constant acceleration 4 m/s^2 has an initial velocity of 0. What is its velocity after it travels 2 m?

McDermott & P.E.G., U.Wash./ *Physics by Inquiry*

Sample of an incorrect solution

$$v_f - v_i \; = \; 2a\Delta x$$

$$v_f - 0 \; = \; 2(4 \text{ m/s}^2)(2 \text{ m})$$

$$v_f \; = \; 16 \text{ m}^2/\text{s}^2$$

Sample of error detection by unit checking

The answer should have units of m/s since the answer is a velocity. We got m²/s², so something is wrong. Checking the units on the first line of the solution, we find:

$$\text{m/s} - \text{m/s} = (\text{m/s}^2)(\text{m})$$

All terms in an equation should have the same units. However, we have m/s, m/s, and m²/s². We know therefore that something is wrong. Indeed, the equation should be:

$$v_f^2 - v_i^2 = 2a\Delta x,$$

which leads to

$$v_f \; = \; 4 \text{ m/s},$$

which is the correct solution and has the right units. Thus checking the units can be helpful in making sure a formula has been remembered correctly.

Paying attention to the units in a calculation is often called *dimensional analysis.*

Dimensional analysis is helpful in checking the mathematical steps carried out in solving

a problem and also in checking that formulas are written correctly in the first place.

Exercise 16.2

Check the units in the following problem solution. If there is an error, correct it and check the units again.

Problem: An object has uniform acceleration. Its velocity is 4 m/s at the start of a 3 m track and twice that at the end. Find Δt for the interval.

Proposed solution:

$$\frac{\Delta x}{\Delta t} \; = \; \frac{v_i + v_f}{2}$$

$$\Delta t \; = \; \frac{v_i + v_f}{2\Delta x}$$

$$\Delta t \; = \; \frac{8 \text{ m/s} + 4 \text{ m/s}}{6 \text{ m}}$$

$$\Delta t \; = \; 2 \text{ m/s}$$

Sample problem

A driver in a car with initial velocity 30 m/s sees a telephone pole ahead and applies the brakes. When the brakes are applied, the car moves with constant acceleration 6 m/s^2. From the place the brakes are applied, the car slides 72 m and strikes the telephone pole. How much time passes after the brakes are applied until the collision with the telephone pole?

Sample solution

This problem can be solved using equation 16.1:

$$\Delta x = v_i \Delta t + \frac{1}{2} a (\Delta t)^2$$

Let the positive direction be the direction of the car's motion. Then $x = +72$ m, $v = +30$ m/s, and $a = -6$ m/s^2. (The acceleration is negative because it is making the velocity a smaller positive number.) Putting these values into Equation 16.1, we get:

$$72 \text{ m} = 30 \text{ m/s } (\Delta t) + \frac{1}{2} (-6 \text{ m/s}^2)(\Delta t)^2$$

Using the quadratic formula to solve for Δt, we get $\Delta t = 6$ and $\Delta t = 4$ seconds.

Why are there two solutions? The equation we used describes motion in which the acceleration is constant forever. If this were the case, the car would reach the telephone pole twice—once on its way out, and once on its way back after it had turned around. However, the equation only applies until the collision. Thus the only relevant solution is $\Delta t = 4$ seconds. The situation is shown on the following graph: the solid line represents the actual motion and the dashed line shows the hypothetical continued motion after the car has reached the telephone pole.

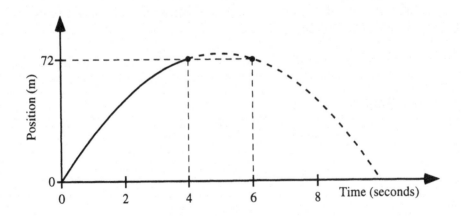

McDermott & P.E.G., U.Wash./ *Physics by Inquiry*

Sample format

Given: $\Delta x = 72$ m

$v_i = 30$ m/s

$a = -6$ m/s^2

Find: Δt

$$\Delta x = v_i \Delta t + \frac{1}{2} a (\Delta t)^2$$

$$72 \text{ m} = 30 \text{ m/s } \Delta t - 3 \text{ m/s}^2 (\Delta t)^2$$

$$(\Delta t)^2 - 10 \text{ s } \Delta t + 24 \text{ s}^2 = 0$$

$$\Delta t = \frac{10 + \sqrt{10^2 - (4)(1)(24)}}{2} \text{ seconds}$$

Answer: $\boxed{\Delta t = 6 \text{ seconds or 4 seconds, but only } \Delta t = 4 \text{ seconds is meaningful}}$

Sample problem

A car with constant acceleration was initially at rest. It traveled 75 m in 5 seconds. What was its final speed 5 seconds after it started?

Sample solution

In our formulas (16.1–16.4), the final velocity is related to other quantities in three equations:

$$v_f = v_i + a \Delta t$$

$$v_f^2 - v_i^2 = 2a \Delta t$$

$$\frac{\Delta x}{\Delta t} = \frac{1}{2} (v_i + v_f)$$

The first will not work. From the phrase "initially at rest," we know that the initial velocity, v_i, is zero. We also know that Δx is 75 m. But we do not know a. The second equation will not work either, again because we do not know a. In the third equation, however, we know the value of every quantity except the final velocity.

Substituting into equation 16.3, we have:

$$\frac{75 \text{ m}}{5 \text{ s}} = \frac{1}{2}(0 + v_f)$$

$$30 \text{ m/s} = v_f$$

Sample format

Given: $\Delta x = 75 \text{ m}$

$\Delta t = 5 \text{ seconds}$

$v_i = 0$

Find: v_f

$$\frac{\Delta x}{\Delta t} = \frac{v_i + v_f}{2}$$

$$\frac{75 \text{ m}}{5 \text{ s}} = \frac{0 + v_f}{2}$$

Answer: $\boxed{30 \text{ m/s} = v_f}$

Sample problem

An object travels 21 meters in an interval of 3 seconds and has velocity 10 m/s at the end of the interval. The motion is uniformly accelerated. Find the acceleration.

Sample solution

Three of our equations contain *a:*

$$\Delta x = v_i \Delta t + \frac{1}{2}a(\Delta t)^2$$

$$v_f = v_i + a\Delta t$$

$$v_f^2 - v_i^2 = 2a\Delta x$$

All of these equations involve the initial velocity, which we do not know. We must therefore find the initial velocity as an intermediate step. The only equation that does not contain the acceleration is:

$$\frac{\Delta x}{\Delta t} = \frac{1}{2}(v_f + v_i)$$

Substituting $x = 21$ m, $\Delta t = 3$ seconds, $v_f = 10$ m/s into this equation, we get:

$$\frac{21 \text{ m}}{3 \text{ s}} = \frac{1}{2}\,(v_i + 10 \text{ m/s})$$

$$4 \text{ m/s} = v_i$$

We can now find the acceleration using any of three equations that contain it. The easiest is probably:

$$v_f = v_i + a\Delta t$$

Substituting into this equation we get:

$$10 \text{ m/s} = 4 \text{ m/s} + a\,(3 \text{ s})$$

$$2 \text{ m/s}^2 = a$$

Sample format

Given:

$\Delta x = 21$ m

$\Delta t = 3$ seconds

$v_f = 10$ m/s

Find:

a

$$\frac{\Delta x}{\Delta t} = \frac{v_i + v_f}{2}$$

$$\frac{21 \text{ m}}{3 \text{ s}} = \frac{v_i + 10 \text{ m/s}}{2}$$

$$4 \text{ m/s} = v_i$$

$$v_f = v_i + a\Delta t$$

$$10 \text{ m/s} = 4 \text{ m/s} + a(3 \text{ s})$$

Answer:

$$\boxed{2 \text{ m/s}^2 = a}$$

Sample problem

A ball is thrown upward with a speed of 30 m/s. How high will it go? The free fall acceleration is 10 m/s^2.

Sample solution

We are not told the amount of time the ball will be moving upward, so we may have to find Δt as an intermediate step. We can find Δt from Equation 16.2:

$$v_f = v_i + a\Delta t$$

First we must decide which way we want to be positive. If we let upward be positive, then v_i is positive and Δx is positive. The acceleration is making the velocity a smaller positive number, so the acceleration is negative. At the top of the ball's flight, the velocity is changing from positive to negative, so it must be zero: $v_f = 0$. Then equation 16.2 becomes:

$$0 = 30 \text{ m/s} + (-10 \text{ m/s}^2)\Delta t$$

$$\Delta t = 3 \text{ seconds}$$

We can now find how high the ball goes from equation 16.1:

$$\Delta x = v_i \Delta t + \frac{1}{2} a(\Delta t)^2$$

$$\Delta x = 30 \text{ m/s } (3 \text{ s}) - 5 \text{ m/s}^2 (3 \text{ s})^2$$

$$\Delta x = 45 \text{ m}$$

In addition to the previous method, there is also another way to solve the problem. We can obtain the answer directly from Equation 16.4 as shown below.

$$v_f^2 - v_i^2 = 2a\Delta x$$

$$(0)^2 - (30 \text{ m/s})^2 = 2(-10 \text{ m/s}^2)\Delta x$$

$$45 \text{ m} = \Delta x$$

Using this second method, we do not ever need to know Δt.

Comment

This problem and the preceding sample problem illustrate a feature of constant acceleration problems that must always be kept in mind. If an important piece of information is not given, either it is not needed at all (as in the last problem) or it must be calculated as an intermediate step (as in the next to last problem).

Sample format

Given: $v_i = 30$ m/s

$v_f = 0$

$a = 10$ m/s^2

Find: Δx

$$v_f^2 - v_i^2 = 2a\Delta x$$

$$(0)^2 - (30 \text{ m/s})^2 = 2(-10 \text{ m/s}^2)\Delta x$$

Answer: $\boxed{\Delta x = 45 \text{ m}}$

Exercise 16.3

A car is slowing down with constant acceleration. When we first see it, its speed is 30 m/s, but every second that goes by its speed drops by 4 m/s.

A. How fast will the car be going 5 seconds after we first see it?

B. How far will the car have traveled during that 5 seconds?

✔ Check your reasoning with a staff member.

Exercise 16.4

Suppose an object obeys the following equation: $x = 10t - 5t^2$

A. Where is the object at $t = 0$?

B. Where is the object at $t = 2$?

C. Where is the object at $t = 4$?

D. Describe the entire motion from $t = 0$ to $t = 4$.

Supplementary problems for *Kinematics*

Note: Many of the following problems ask that you do not use algebra in your solution. For those problems, do not base your answer on an algebraic formula. Explain your reasoning based on your interpretations of the quantities involved.

Problem 1.1

Suppose an object moves with uniform motion along a track with a definite start and finish. Its velocity is 84 cm/s. At 3.6 seconds after the start, the object still has 156 cm to go before reaching the finish line of the track.

What is the total time interval from start to finish?

Explain your reasoning. Do not use algebra.

Problem 1.2

A. An object moving with uniform motion traveled 265 cm in 2.4 seconds. How much time did the object take to travel the last 50 cm? Explain your reasoning. Do not use algebra.

B. Make up and solve an analogous problem concerning the mass and volume of some kind of metal. What is given, what is asked, and the reasoning used should all be related to each other exactly as in part A.

C. Identify all the corresponding quantities of the two problems and tell how they correspond to each other.

Problem 1.3

On many interstate highways, numbered posts are placed every mile alongside the road. At right is a record of the clock times at which a car passed several mileposts.

Mile number	Clock time
40	1:00
50	1:11
60	1:22
70	1:33

A. Plot a graph of mile number versus clock time. The origin of the graph (the point where the axes cross) should be mile number 0 and clock time 1:00. *Use graph paper!*

B. Is the motion uniform?

C. What is the slope of this graph at each of the measured clock times?

Problem 1.4

At right is a graph of part of the motion of a uniformly moving object.

A. What does just the one point *P* tell you about the motion?

B. What does the length *a* represent?

C. What does the length *b* represent?

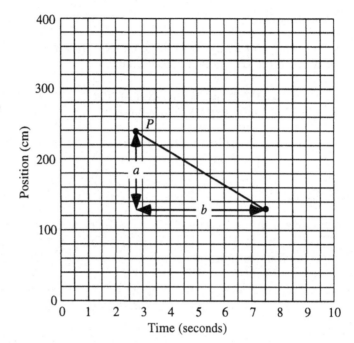

Problem 1.5

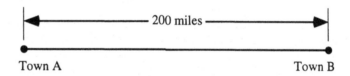

Two towns, A and B, are 200 miles apart. Car 1 starts from town A and, at the same time, car 2 starts from town B. They move toward each other with constant speeds: car 1 at 50 mph, car 2 at 40 mph.

A. How far apart are the two cars if car 2 has traveled a time Δt?

B. Where do the cars meet? If you use an equation, interpret the expression on each side and tell why you set them equal.

Problem 1.6

Two cars travel at the same constant speed. The first car travels for 10 minutes, the second for 27 minutes. If the second car travels 18 miles farther than the first, how far does each car travel? Solve algebraically.

Problem 1.7

Two cars start from the same place at the same time. After two hours, one car has traveled 30% farther than the other. If one car's speed is 20 mph larger than the other car's speed, what are the speeds of the cars? Solve algebraically.

Problem 1.8

A bug is 10 feet away from the base of a tree at noon. It is creeping slowly but steadily away from the tree at a constant speed v.

A. Write an algebraic expression for the bug's distance from the tree at time t, where t is the time that has passed since noon.

B. If the bug is 30 feet from the tree at 12:50 P.M., when will it be 60 feet from the tree?

Problem 1.9

A traveler left home on a trip across the desert. He took along enough provisions for a 19-day journey. He is able to travel with a constant speed s. After 15 days, he is still 100 km from his destination.

Write an expression for the number of days of provisions he will have left when he arrives. Explain your reasoning in detail.

Can your expression ever be negative? What would that mean and why?

Problem 1.10

Two objects start 30 km apart at the same time and move toward each other with uniform motion. Object A has a speed of 6.3 km/h and object B has a speed of 4.5 km/h. When will they meet?

Problem 1.11

A car and a truck are at a gas station. At 2:00 the truck starts traveling along the highway at a constant speed of 54 mph. A quarter of an hour later, at 2:15, the car starts traveling along the same highway with a constant speed of 60 mph. At what time will the car catch up with the truck?

Solve this problem algebraically using the following steps.

A. Let Δt be the number of hours the car travels between leaving the gas station and catching up with the truck. Is Δt the answer? If not, how can you get the answer once you find Δt?

B. Write an expression for the car's distance from the gas station Δt hours after the car starts. Explain your reasoning.

C. Write an expression for the truck's distance from the gas station Δt hours after the car starts. Explain your reasoning.

D. Write your equation and tell why you set the two sides equal.

E. Solve your equation and answer the question.

Problem 1.12

Two cars start at the same time from the same place and travel along the same road. Car A has a constant speed of 50 mph. Car B travels with constant speed v for 100 miles, then stops to rest for a time t, then travels again with constant speed v for another 100 miles. Assume v is greater than 50 mph.

A. Write algebraic expressions as answers to the following questions:

   (1) How far is B ahead of A when B stops to rest?

   (2) Where is A when B starts up again?

   (3) What time is it when A passes a marker 150 miles from the starting point?

   (4) Suppose we were at the 150-mile marker. How long would we have to wait between the passing of A and the passing of B?

   (5) Where is A when B arrives at a point 200 miles from its starting point?

B. Suppose A and B arrive at the 200-mile position at the same time.

   (1) Write an equation that states this condition.

   (2) How long does B rest if its speed is 60 mph?

   (3) What must B's speed be if it rests for one hour?

C. Plot a position versus time graph for the motion of both cars on the same axes for the motions described in part B(2) above.

Problem 1.13

Two cars start from the same place on the same road but at different times. Car 1 travels with constant velocity v_1 for a distance d, and then stops. Car 2 starts at time t after car 1, travels with constant velocity v_2 until it has gone the same distance d, and then stops.

Give a verbal interpretation of each of the following expressions, if such an interpretation exists. (Some expressions may have no interpretation relevant to the motions described.)

A. d/v_2 B. $v_1 t$ C. d/t D. $v_2 t$

      McDermott & P.E.G., U.Wash./ *Physics by Inquiry*

Problem 1.14

Two objects move in the same direction. Both have uniform motion. Object 1 starts at position 0 at time 0. Object 2 starts at position a at time b. They meet at position c at time d.

Give verbal interpretations of the following expressions, if such an interpretation exists. (Some of the expressions have no relevant interpretation.)

      A. a/b D. $(c - a)/(d - b)$

      B. c/b E. $(a - b)/(c - d)$

      C. c/d

Problem 2.1

Tell whether the following statements are *always* true. If the statement is not always true, give an example for which it is false. In either case explain your reasoning.

A. If two objects both reach the same position at the same time, then they must have the same speed at that one instant.

B. On the freeway, if two cars reach the same speed, then they must be side by side.

C. Two objects are traveling on parallel straight tracks. If the objects are getting closer together, then the objects must be moving in opposite directions.

Problem 2.2

Two cars are traveling from point A to point C. Each car moves with uniform motion throughout the entire trip.

Point B is 25 km from point A, and point C is 40 km beyond point B as shown.

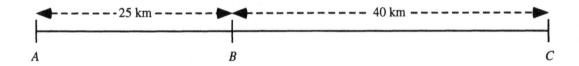

Car 1 leaves point A at 12:00 and reaches point B at 12:20. Car 2 leaves point A at 12:10 and reaches point B at 12:27.

Which car will arrive at point C first? Explain your reasoning thoroughly.

Problem 2.3

Make an analogy between a position along a line and an instant in time. Include the similarity between displacement and duration. Identify corresponding parts and tell how they are similar.

Problem 2.4

Comment on the following reasoning. Is it correct or incorrect? If it is correct, explain the reasoning. If it is incorrect, point out what is wrong with it.

"Velocity is m/s, x is meters, and t is seconds, so v = x/t."

Problem 2.5

The following is a position versus time graph for the motions of two objects, A and B, that are moving along the same meter stick.

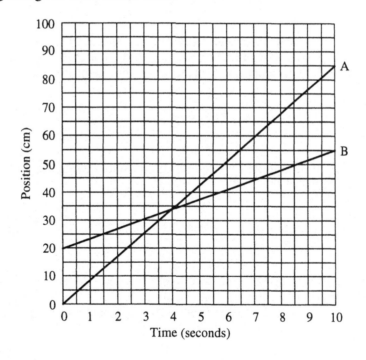

A. At the instant $t = 2$ seconds, is the speed of object A greater than, less than, or equal to the speed of object B? Explain your reasoning.

B. Do objects A and B ever have the same speed? If so, at what times? Explain your reasoning.

McDermott & P.E.G., U.Wash./ *Physics by Inquiry*

Problem 2.6

One way people study volcanic eruptions is to measure the distribution of ash that falls from the air. Generally, most of the ash falls close to the volcano, though some ash may fall hundreds of kilometers away. A quantity that is useful in describing the distribution is the mass of ash per square meter that fell at different places after an eruption.

Make an analogy between the concepts of instantaneous velocity and the mass of ash per square meter. Identify what corresponds to each of the following and tell how they are alike.

A. instantaneous velocity

B. displacement, Δx

C. duration, Δt

D. the fact that instantaneous velocity may be different at different times

Problem 2.7

Make an analogy between instantaneous velocity and a quantity we have not studied in this course. Identify the quantities that correspond to displacement, duration, and velocity, and tell how they correspond. Define all of the quantities you use.

Problem 2.8

A boat travels 6 miles in 35 minutes. A second boat travels twice the distance in half the time. What are the speeds of the boats?

Problem 2.9

Ports A and B are 400 miles apart. One boat starts from port A for port B at 6:05 and travels with constant speed 12 mph. If it leaves port A at 8:20, what speed must a second boat have to arrive at port B at the same time? Explain your reasoning.

Problem 5.1

On a straight inclined track, a ball rolling uphill takes 1.3 seconds to slow down from a velocity of 135 cm/s to a velocity of 65 cm/s. On the same track, how much time would it take a rolling ball with velocity 85 cm/s to reach a velocity of 0? Explain your reasoning. Do not use algebra.

Problem 5.2

Suppose an oil tanker is out of control and is coasting to a stop with constant acceleration. At 1:15 its velocity is 20 mph. At 1:40 its velocity is 12 mph. When will it stop? Explain your reasoning. Do not use algebra.

Problem 5.3

An object with initial velocity 14 m/s moves with constant acceleration. Its velocity increases by an amount Δv in 5 seconds.

A. Write an expression for the final velocity.

B. Write an expression for the acceleration.

Problem 5.4

What is the initial velocity of an object with acceleration 5 mph/s if it triples its velocity in 4 seconds? Solve algebraically.

Problem 5.5

Object A has initial velocity 18 mph, and object B has initial velocity 6 mph. If A's acceleration is 0.06 mph/s and B's is 0.10 mph/s, how long must they accelerate before they have the same velocity?

Problem 5.6

Tell whether each of the following statements is *always* true. If the statement is not always true, give an example in which it is not true. In either case, explain your reasoning.

A. Two objects are traveling on parallel straight tracks. If the distance between them is getting larger, then at least one of the objects must have a nonuniform motion.

B. In a foot race, if runner A is catching up to runner B, then runner A *must* be speeding up.

C. If an object moving on a straight track does not have uniform motion, then it must be slowing down or speeding up.

Problem 5.7

This problem concerns the following dialog:

Student 1: *"Acceleration is m/s². Meters is Δx and seconds is Δt, so*
a = Δx/(Δt)²."

Student 2: *"No, acceleration is really m/s/s. Meters is Δx and seconds is Δt,*
so therefore a = (Δx/Δt)/Δt."

A. Are these students saying essentially the same thing, or is there an important difference between their statements?

B. Both students are reasoning incorrectly. Explain why they are wrong.

Problem 6.1

In the following problem, sketch the v versus t graph that corresponds to the x versus t graph given. Do not actually compute the values for the v versus t graph. Just indicate the general shape of the graph. Label times A through F on your graph.

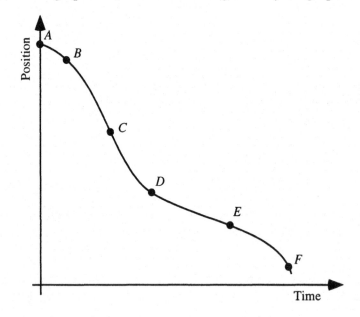

          McDermott & P.E.G., U.Wash./ *Physics by Inquiry*

Problem 6.2

Accurately plot the v versus t graph for the motion represented by the x versus t graph below. Plot points every 10 seconds.

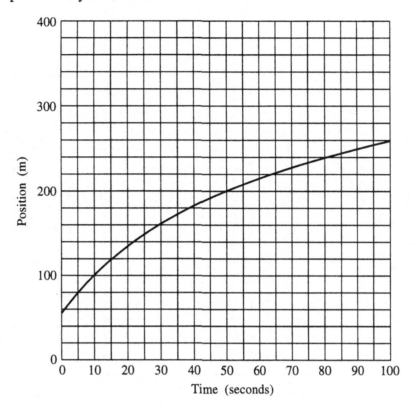

Problem 6.3

In the following problem, sketch the v versus t graph that corresponds to the x versus t graph given. Do not actually compute the values for the v versus t graph. Just indicate the general shape of the graph. Label times A through F on your graph.

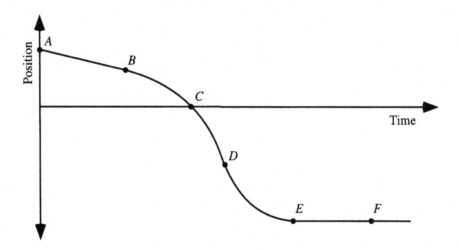

McDermott & P.E.G., U.Wash./ *Physics by Inquiry*

Problem 6.4

Accurately plot a *v* versus *t* graph for the motion represented by the *x* versus *t* graph below. Plot points every half second.

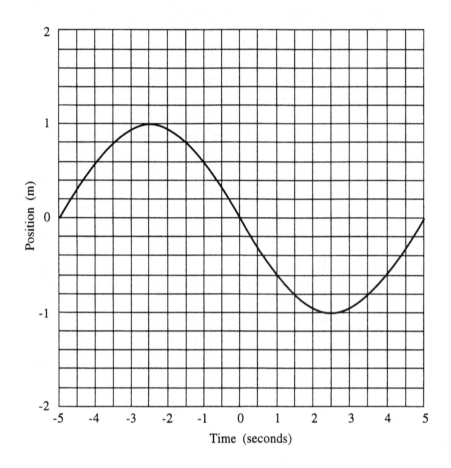

Problem 6.5

Sketch the *v* versus *t* graph for the motion represented below. Label times *A, B, C, D,* and *E* on your graph.

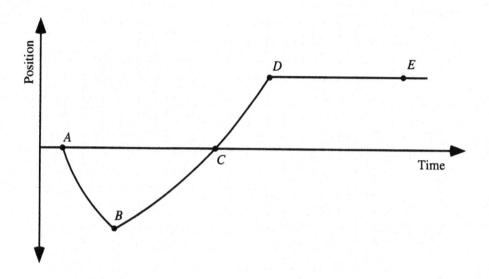

Problem 6.6

Sketch the *v* versus *t* graph for the motion represented below. Label points *A, B,* and *C* on your graph.

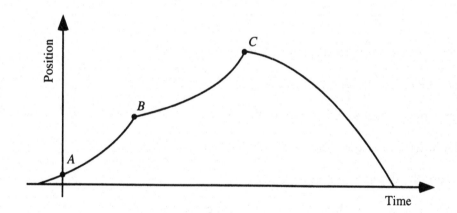

McDermott & P.E.G., U.Wash./*Physics by Inquiry*

Problem 7.1

Answer the following questions based on the graph below. Explain your reasoning.

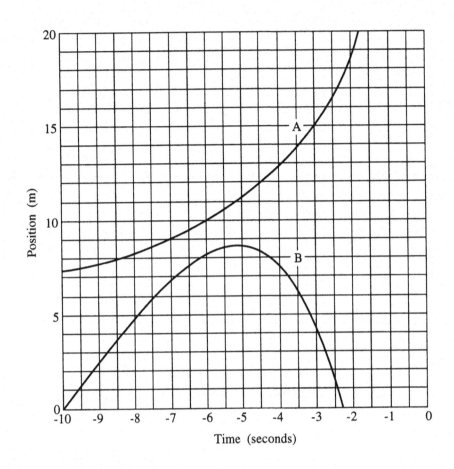

A. Which object has the greater speed at $t = -9$ seconds?

B. Which object has the greater speed at $t = -3$ seconds?

C. Are the objects ever side by side (at the same position)? If so, when?

D. Do the objects ever have the same velocity? If so, when?

E. When are the objects closest together?

F. How far apart are the objects when they are closest together?

McDermott & P.E.G., U.Wash./ *Physics by Inquiry*

Problem 7.2

The questions below apply to the following graph.

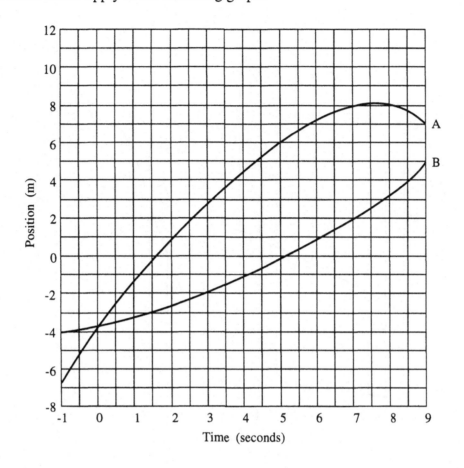

A. At $t = 0$, is object A speeding up or slowing down?

B. At $t = 0$, is object B speeding up or slowing down?

C. At $t = 0$, which object is passing the other?

D. Do the objects ever have the same velocity? If so, when?

E. At what times, if any, does object A turn around?

F. At what times, if any, does object B turn around?

McDermott & P.E.G., U.Wash./ *Physics by Inquiry*

Problem 7.3

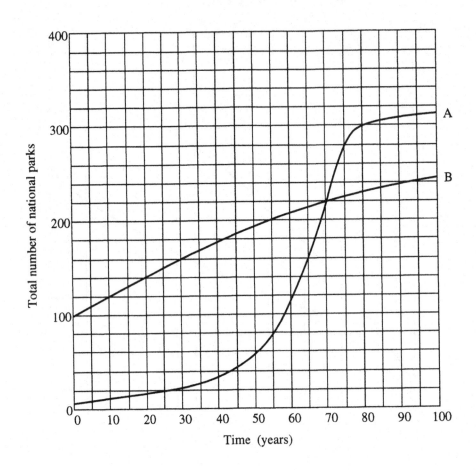

The following questions apply to the graph above. This graph describes the history of national park creation in two countries, A and B. The graph shows how many parks there were in each country at any date over the period of a century.

A. Were the countries ever creating the same number of parks per year? If so, when?

B. At $t = 50$, which country was creating more parks per year?

C. Just before $t = 70$, which country was creating more parks per year?

D. At $t = 90$, which country was creating more parks per year?

Problem 8.1

Draw a diagram of a set of tracks you could use to produce the motion represented on the graph below. Describe the motion in words.

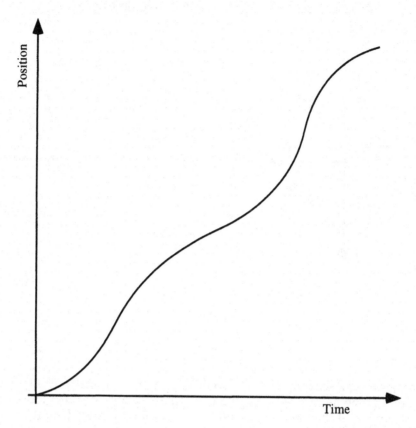

McDermott & P.E.G., U.Wash./*Physics by Inquiry*

Problem 8.2

A. Draw a diagram of a set of tracks you could use to produce the motion represented in the *v* versus *t* graph below. Also sketch an *x* versus *t* graph for this motion. Give a verbal description of this motion and what you would do to produce it.

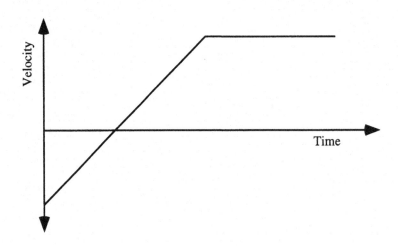

B. Draw a diagram of a set of tracks you could use to produce the motion represented in the *x* versus *t* graph below. Also sketch a *v* versus *t* graph for this motion. Give a verbal description of this motion and what you would do to produce it.

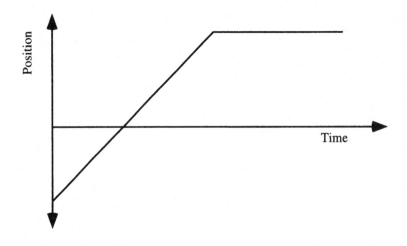

McDermott & P.E.G., U.Wash./ *Physics by Inquiry*

Problem 9.1

The clerk at a silver mine kept a record of the total profit made by the mine since it opened. In the beginning, the total profit increased rapidly. As the silver became less concentrated, however, the total profit stopped rising so quickly and later began to fall. Eventually, the total profit became negative, indicating an overall loss. The mine owners eventually shut the mine after a substantial loss, despite warnings from the mine clerk.

When a certain amount of silver has already been mined, the profit on the *next* ton is called the *marginal profit*. For example, suppose 18 tons of silver has already been mined giving a total profit of $2,000,000 so far. If the next ton mined yields a profit of $200,000, the marginal profit would be $200,000/ton at this stage of mine operations. Thus, after 19 tons has been mined, the total profit would be $2,200,000. Later on, when the silver is harder to get, the marginal profit might well be lower.

Below is a graph of the total profit versus the amount of silver already mined.

    A. At which point was the marginal profit maximum? How can you tell?

    B. At which point did the marginal profit first become negative? What does the negative sign mean?

    C. Sketch a graph of the marginal profit versus the amount of silver already mined. Label points corresponding to A through G on your graph.

    D. At which point would you have shut the mine and why?

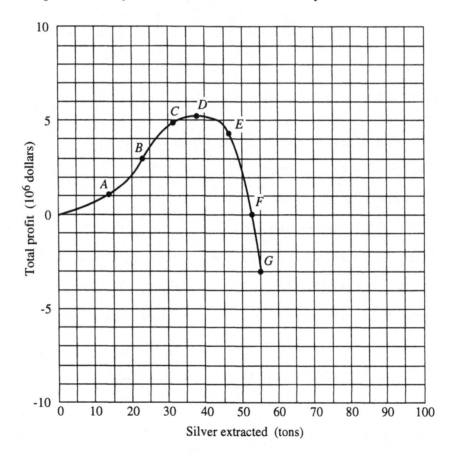

Problem 9.2

Below is a graph of population versus time for a hypothetical country during a period of famine.

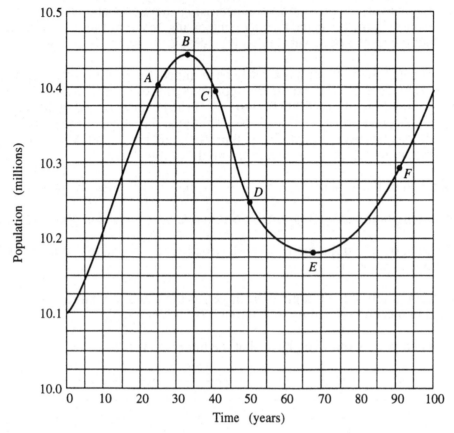

A. List all of the lettered points, if any, at which each of the following is true. Explain your reasoning.

(1)   The population is neither increasing nor decreasing.

(2)   The population is maximum.

(3)   The population growth rate is zero.

(4)   The population is increasing and the population growth rate is increasing.

(5)   The population is increasing and the population growth rate is decreasing.

(6)   The population is declining and the rate of population decline is decreasing.

(7)   The population is declining and the rate of population decline is increasing.

(8)   The population is increasing more and more slowly.

(9)   The population is decreasing more and more quickly.

(10)   The population is decreasing more and more slowly.

B. Sketch a graph of the population growth rate versus time.

McDermott & P.E.G., U.Wash./ *Physics by Inquiry*

Problem 9.3

The graph below shows the variation in the number of head of cattle over time in the republic of Forwichistan.

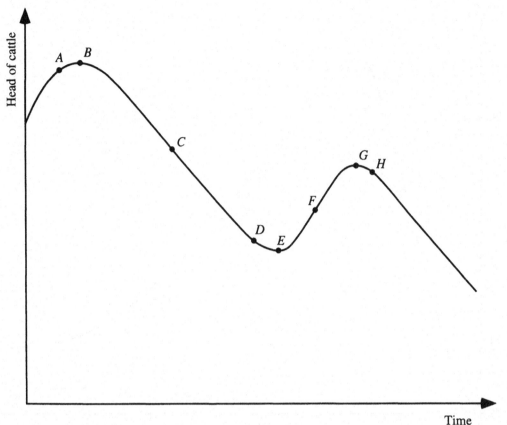

A. List all of the lettered points, if any, at which each of the following is true. Explain your reasoning.

(1) The number of cattle is minimum.

(2) The number of cattle is maximum.

(3) The rate of decrease of cattle is neither increasing nor decreasing.

(4) The rate of increase of cattle is maximum.

(5) The rate of change in the number of cattle is zero.

(6) The number of cattle is increasing and the rate of increase of cattle is decreasing.

(7) The number of cattle is decreasing and the rate of decrease of cattle is diminishing.

(8) The number of cattle is decreasing more and more slowly.

(9) The number of cattle is decreasing more and more quickly.

(10) The number of cattle is decreasing the fastest.

B. Sketch a graph of the rate of change of number of cattle versus time.

     McDermott & P.E.G., U.Wash./ *Physics by Inquiry*

Problem 9.4

The following graph shows the variation of temperature with depth of water in a lagoon that is popular among tourists in Forwichistan.

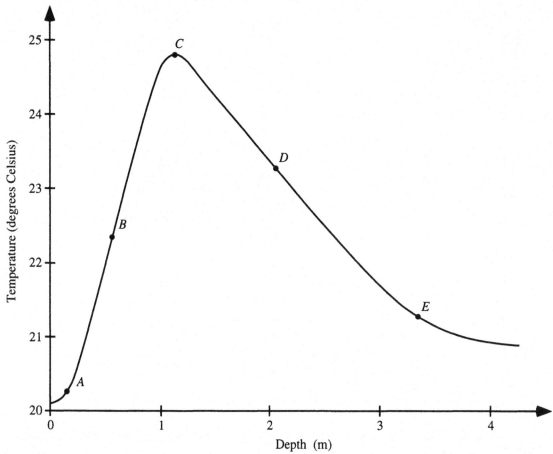

The temperature gradient is a quantity that tells how much the temperature changes per unit depth.

A. List all of the lettered points, if any, at which each of the following is true.

   (1)   The temperature is increasing the fastest with depth.

   (2)   The temperature is maximum.

   (3)   The temperature gradient is maximum.

   (4)   The temperature gradient is 0.

   (5)   The temperature and the temperature gradient are both increasing.

   (6)   The temperature and the temperature gradient are both decreasing.

   (7)   The temperature is increasing more and more quickly.

   (8)   The temperature is decreasing more and more slowly.

B. Sketch a graph of temperature gradient versus time.

                    McDermott & P.E.G., U.Wash./ *Physics by Inquiry*

Problem 10.1

The questions below apply to the v versus t graph that follows. The objects described move along parallel tracks.

A. Was the acceleration of A greater than, less than, or equal to the acceleration of B

  (1) at $t = 0$ second?

  (2) at $t = 1.5$ seconds?

  (3) at $t = 4$ seconds?

B. Are the signs of the acceleration of A and B the same or different at $t = 1$ second?

C. At what times, if any, are the objects side by side?

D. What happens at $t = 1.5$ seconds?

E. What happens at $t = 4$ seconds?

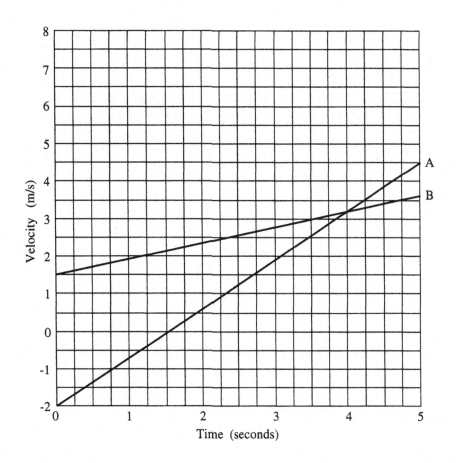

Problem 10.2

A certain hamburger company keeps a running tally of the total number of hamburgers it has ever sold. Here is a fictitious interview with a company representative:

Interviewer:	*"What are your annual sales?"*
Representative:	*"1.62 billion, currently."*
Interviewer:	*"What are your prospects for the future?"*
Representative:	*"Excellent. Our annual sales are increasing 0.03 billion per month as of now."*

Make an analogy between the quantities discussed above and other quantities involved with the hamburger business and the quantities from motion listed below. Tell what corresponds to each and tell how they are alike.

A. acceleration

B. velocity

C. displacement

D. duration

E. time

F. position

Problem 11.1

In each of the following, a problem and a solution are given. Tell whether the solution is correct or incorrect. If it is correct, tell why. If it is incorrect, tell why.

A. *Problem:* A solution contains 15 g of solute dissolved in water. If the solution is diluted so that its volume is 600 mL, how much solute will it contain?

Proposed solution:

$$\frac{M_1}{V_1} = \frac{M_2}{V_2}$$

$$\frac{15}{400} = \frac{M_2}{600}$$

$$M_2 = 22.5 \text{ g}$$

B. *Problem:* Two runners ran a race between two trees. The first runner had a constant speed of 8.0 m/s and took 50.2 seconds. If the second runner had a constant speed of 7.8 m/s, what was his time?

Proposed solution:

$$v_1 \Delta t_1 = v_2 \Delta t_2$$

$$8.0(50.2) = 7.8 \Delta t_2$$

$$\Delta t_2 = 51.5 \text{ seconds}$$

McDermott & P.E.G., U.Wash./*Physics by Inquiry*

Problem 11.2

In each of the following, a problem and a solution are given. Tell whether the solution is correct or incorrect. If it is correct, tell why. If it is incorrect, tell why.

A. *Problem:* We have two pieces of metal with the same mass. The first piece has a volume of 54 cm^3 and a density of 2.7 g/cm^3. If the second piece has a volume of 72 cm^3, what is its density?

Proposed solution:

$$\frac{D_1}{V_1} = \frac{D_2}{V_2}$$

$$\frac{2.7}{D_2} = \frac{54}{72}$$

$$(2.7)\frac{72}{54} = D_2$$

$$3.6\,\text{g}/\text{cm}^3$$

B. *Problem:* We have a meterstick suspended from a fulcrum at its center. On the right side is a 50 g mass 21 cm from the fulcrum. How much mass must we hang 35 cm from the fulcrum on the left side to balance the stick?

Proposed solution:

$$\frac{M_1}{L_1} = \frac{M_2}{L_2}$$

$$\frac{50}{21} = \frac{M_2}{35}$$

$$(35)\frac{50}{21} = M_2$$

$$83 = M_2$$

          McDermott & P.E.G., U.Wash./*Physics by Inquiry*

Problem 12.1

Find the displacement from $t = 2$ seconds to $t = 8$ seconds for the motion below.

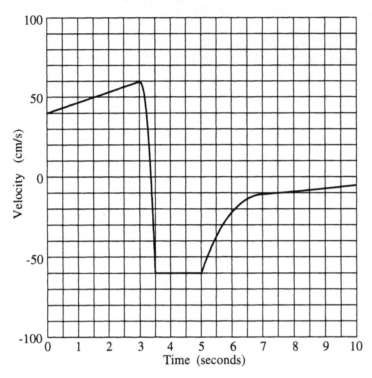

Problem 12.2

An object is located at $x = 0$ cm at $t = 0$ seconds. (See graph below.) How much time does it take for the object to travel from $x = 55$ cm to $x = 350$ cm?

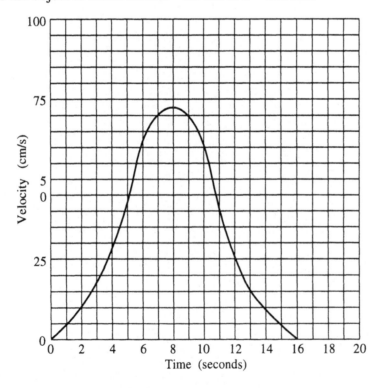

Problem 13.1

Below is a graph of instantaneous acceleration versus time for a moving object.

A. Assume the object starts at rest at $t = 0$ second and plot a v versus t graph for this motion. Plot points every 10 seconds.

B. Assume the object starts at position $x = -10$ m and plot a position versus time graph for this motion. Plot points every 10 seconds.

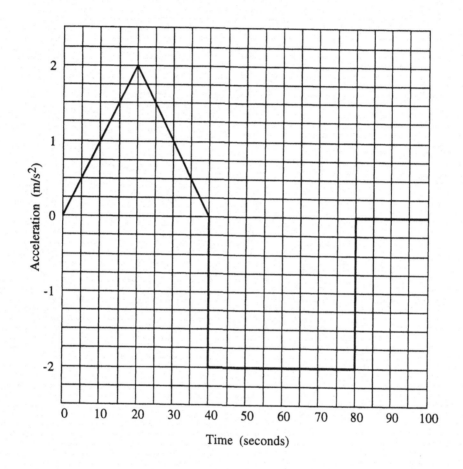

McDermott & P.E.G., U.Wash./ *Physics by Inquiry*

Problem 13.2

Below is a graph of housing growth rate versus time for a hypothetical city during an age of urban decay. The number of single-family housing units at the beginning of the interval shown was 500,000. Plot a graph of the number of single-family housing units versus time for the interval shown. Plot points every half year.

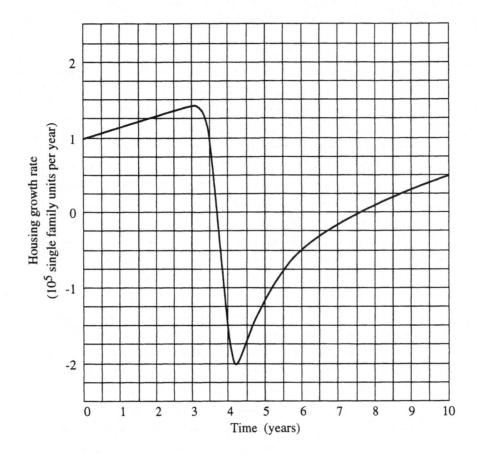

Problem 13.3

Below is a graph of the population growth rate of a certain city versus time for a century. The population of the city was 4,200,000 in 1830.

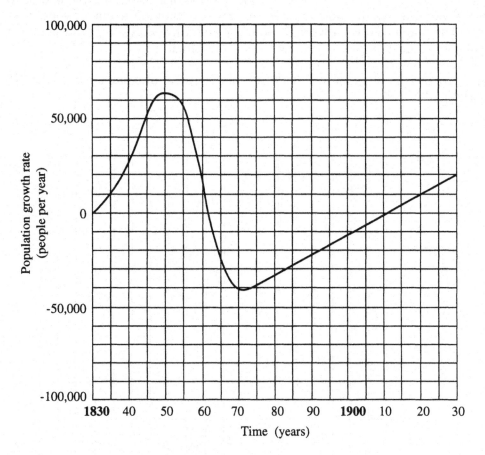

A. When (what year) was the population of the city the largest?

B. When was the population of the city the smallest?

C. When was the population of the city increasing the fastest?

D. When was the population of the city decreasing the fastest?

E. When was the population of the city changing the slowest?

F. What was the largest population the city had in the century shown?

G. What was the smallest population the city had in the century shown?

McDermott & P.E.G., U.Wash./ *Physics by Inquiry*

Problem 13.4

A canal carries water for irrigation a great distance over varied terrain and through channels of different sands. Some parts of the canal are exposed to the sun and wind; others are covered by trees. The canal is wide at some places and narrow at others. Because of variations in the water course, there is variation in the rate of water loss along the canal. The graph below shows the amount of water lost daily from the canal per 100 m plotted against distance from the start of the canal.

Find the total amount of water lost daily from the segment of canal between the 10th kilometer and the 100th.

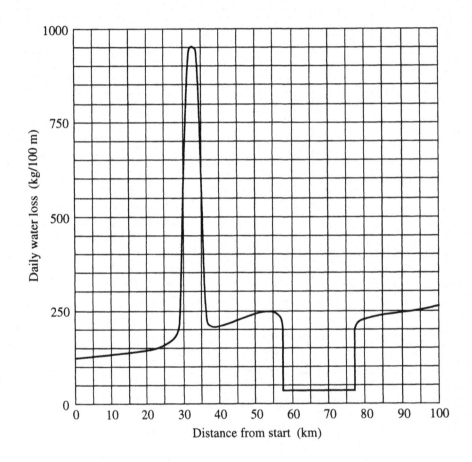

Problem 14.1

Find the average velocity between $t = 0$ second and $t = 10$ seconds for the motion below.

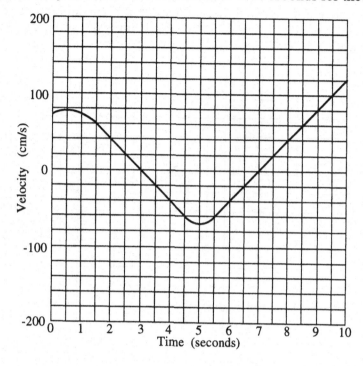

Problem 14.2

A. On the velocity versus time graph below, draw a line representing the average velocity of this motion between $t = 0$ second and $t = 8$ seconds.

B. If the displacement of the object was 22 m between $t = 0$ second and $t = 8$ seconds, what is the scale of the graph?

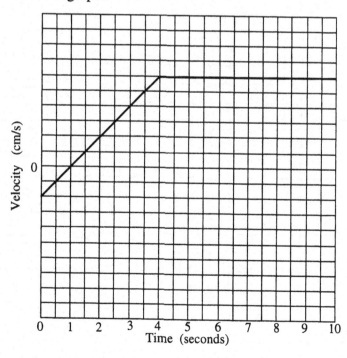

McDermott & P.E.G., U.Wash./ *Physics by Inquiry*

Problem 14.3

This problem is about the unfortunate person who has a backyard vegetable garden near a pesticide factory. The graph below shows the variation in the amount of poison in this person's body over a period of 10 weeks. What is the average amount of poison in the body over the 10 weeks shown?

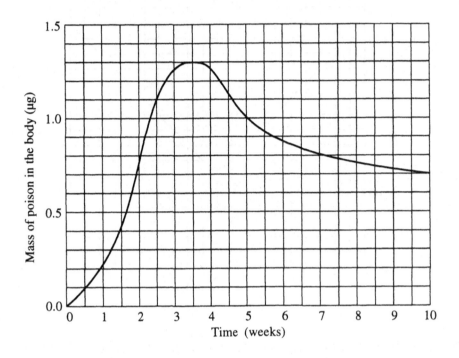

Problem 14.4

Below is the graph of population versus time for a hypothetical country during an age of famine. What was the average population between $t = 50$ years and $t = 90$ years?

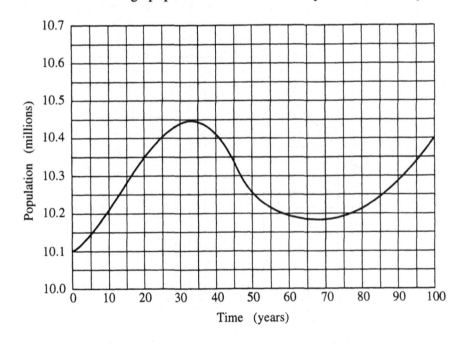

McDermott & P.E.G., U.Wash./ *Physics by Inquiry*

Problem 15.1

The purpose of this problem is to provide practice in deciding how to get information from a graph. The problem consists of six graphs, A through F, with one question to be answered about each.

A. Below is a graph showing the amount of granite that had been excavated from a quarry by each date over a 100-year period. If granite was being excavated at the rate of 27,000 tons each year at $t = 50$ years, how much granite was being excavated each year at $t = 80$ years?

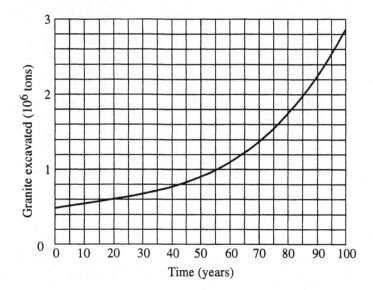

B. Below is a graph of acceleration versus time for a moving object. If the object's velocity is 13 cm/s at $t = 5$ seconds, what is its velocity at $t = 8$ seconds?

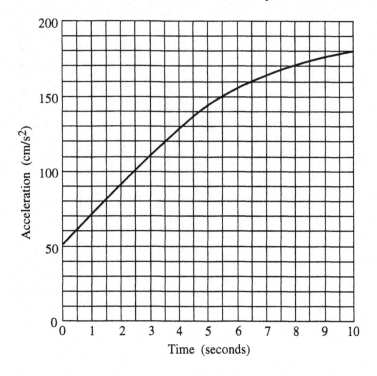

C. Below is a graph showing the rate of growth of a desert over a 100-year period. If the area of the desert described was 270 million km^2 at $t = 50$ years, what was the area at $t = 80$ years?

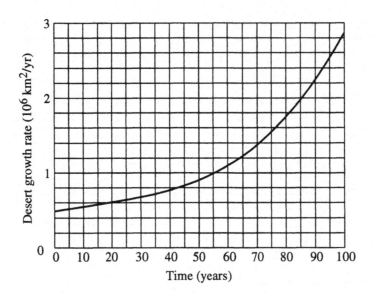

D. Below is a graph of position versus time for a moving object. If the object's velocity is 13 cm/s at $t = 5$ seconds, what is its velocity at $t = 8$ seconds?

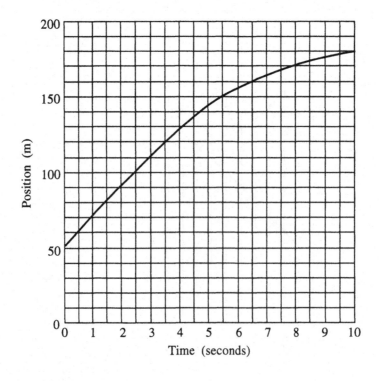

E. A bottle is being filled with water. The graph below shows the volume of water in the bottle from $t = 0$ second until $t = 10$ seconds. If the filling rate is 13 mL/s at $t = 5$ seconds, what is the filling rate at $t = 8$ seconds?

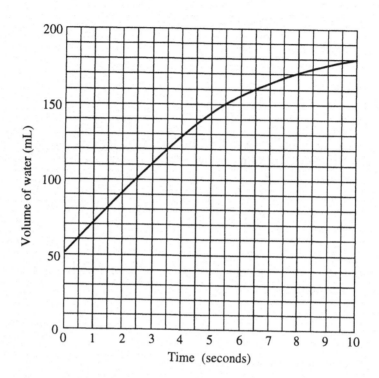

F. Below is a graph showing the annual production of government personnel files over a 100-year period. If the government personnel files contained 270 million sheets at $t = 50$ years, how large were the files at $t = 80$ years?

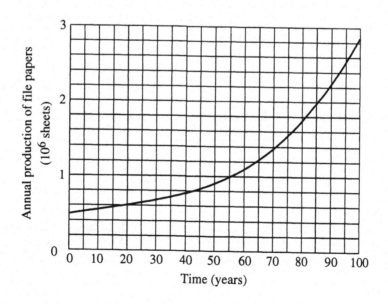

Problem 16.1

An object starts from rest and moves with constant acceleration. After one second, it has traveled 3 m.

A. What is its velocity one second after starting?

B. How far does it travel after two seconds?

Problem 16.2

One object has an acceleration of 15 miles/hour/second. Another object has an acceleration of 15 miles/second/hour. Which of these accelerations is larger, or are they the same?

Problem 16.3

A car slows down with constant acceleration from 50 mph to 30 mph over a distance of 100 feet. What is its acceleration?

Problem 16.4

A train with constant acceleration 0.5 m/s^2 was slowing down. If it traveled 100 m in 2 seconds, what was its initial velocity at the beginning of the 2 seconds?

Problem 16.5

An oil tanker is coasting to a stop and has a constant acceleration of 0.001 mph/s. If its initial speed is 20 mph, how far will it travel before its speed is 10 mph?

Problem 16.6

An airplane taking off attains a take-off speed of 80 m/s after traveling 800 m on the runway with constant acceleration. What is the plane's acceleration?

Problem 16.7

A physicist on the fourth floor throws a water balloon out of his window straight down at the ground. The acceleration of the balloon is 10 m/s^2. If the window is 20 m above the ground and the balloon hits the ground after 0.8 seconds, what is its initial velocity?

Problem 16.8

Suppose a military aircraft flying at tree-top level flies over an enemy surface-to-air missile base at 350 mph. ("Tree-top level" means the plane's height should be ignored.) Realizing he will be fired upon, the pilot accelerates with constant acceleration 15 mph/s to try to get away. At the instant the plane passes the missile base, a missile with constant acceleration 50 mph/s is fired at the plane.

A. How long after firing will the missile have the same speed as the plane?

B. Where will the plane be when the missile strikes it?

Problem 16.9

An object with constant acceleration obeys the following equation:

$$v^2 = 169 + 40\Delta x$$

A. How do you interpret the number *169?*

B. How do you interpret the number *40?*

Problem 16.10

An object with constant acceleration obeys the following equation:

$$16t^2 + 8t = x + 15$$

A. How do you interpret the number *16?*

B. How do you interpret the number *8?*

C. How do you interpret the number *15?*

McDermott & P.E.G., U.Wash./ *Physics by Inquiry*

Problem 16.11

An object with constant acceleration obeys the following equation:

$$x/t = 3.2t + 4.8 - 15/t$$

A. How do you interpret the number *3.2?*

B. How do you interpret the number *4.8?*

C. How do you interpret the number *15?*

Problem 16.12

An object with constant acceleration obeys the following equation:

$$v^2 = 4 + 12t + 9t^2$$

A. What is the initial velocity at $t = 0$?

B. What is the acceleration?

C. What is the initial position at $t = 0$?

PHYSICS BY INQUIRY

Astronomy by Sight:

the earth and the solar system

In *Astronomy by Sight: the sun, moon, and stars,* we developed models to help us predict and explain some of the daily and monthly changes in the appearance of the sky. In *Astronomy by Sight: the earth and the solar system,* we extend the models to account for seasonal changes on the earth and the motion of the planets in the solar system.

Note: *Astronomy by Sight: the earth and the solar system* is the second of two closely related modules. It builds directly on *Astronomy by Sight: the sun, moon, and stars.* Students should finish working through the first module, which appears in Volume I of *Physics by Inquiry,* before beginning the second.

Section 1. The celestial sphere

In *Astronomy by Sight: the sun, moon, and stars,* we have developed models for the sky that were based on extended observations that we have made of the sun, moon, and the stars. In *Astronomy by Sight: the earth and the solar system,* we continue to draw on those observations. As you work through this module you should continue to make shadow plots and observations of patterns of stars and other celestial objects in the sky.

Exercise 1.1

In this exercise we draw on observations that you have made of several stars at the same time of day over a period of several months.

A. Discuss your observations with your partners. At a particular time of day, are the stars always in the same location? If not, describe how the locations of some stars at a fixed time of day appear to change from day to day. Answer for stars in the north, south, east, and west.

Do the stars appear to move through more than, less than, or exactly one complete circle in the sky in 24 hours? Explain.

B. Does the angle between the sun and a particular star stay the same from day to day, or does it change? (*Note:* It is usually not possible to measure the angle between the sun and a star directly. However, you can estimate the location of the sun at a given time during the night.)

C. The figure at right shows several stars that a student sees while looking south at midnight. On the basis of your observations, predict the locations of the stars at midnight several weeks later.

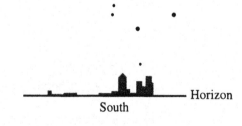

South — Horizon

In Exercise 1.1 you described how the stars change location in the sky from night to night. In the rest of this section, we develop a model that can help us to account for these observations. We begin by imagining that the sun and the stars lie on the surface of a very large sphere with the earth at its center. This imaginary surface is called the *celestial sphere*. In the following experiment, you construct a physical representation of the celestial sphere.

Experiment 1.2

Obtain a celestial sphere kit and masking tape. From the kit, you will need the two transparent half-spheres, the bead representing the earth, one dowel, one drinking straw, and the Styrofoam base.

A. Make a small hole in the center of each half-sphere using a pin or a small nail. Place the earth-bead and straw on the dowel. Cut the straw so that when the half-spheres are joined as shown, the earth-bead rests at the center of the sphere.

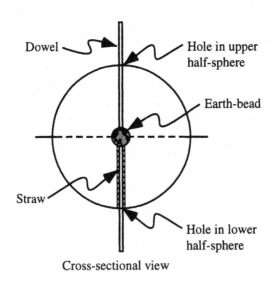

Cross-sectional view

The earth-bead should rotate with the dowel and not slip. Use glue to fasten the earth-bead to the dowel.

Place the dowel and earth-bead inside the half-spheres as shown. Cut the square edge of plastic off the half-spheres. Leave only about 1 cm of plastic. Use tape to hold the half-spheres together.

B. Imagine an observer at your location on the earth-bead.

> Describe each of the following for the observer: the horizon, vertical, north, south, east, west.

> Which portion of the sphere could the observer see?

C. Suppose that the earth-bead were made smaller.

> How would your answers to part B be different?

> Suppose that the earth-bead were made very small relative to the sphere.

>> What portion of the sphere could an observer on the earth see? What would represent the horizon for the observer? Explain.

D. Orient the sphere so that north, south, east, and west for the observer on the earth-bead line up with your north, south, east, and west. Insert the dowel into the Styrofoam base to hold the sphere in place.

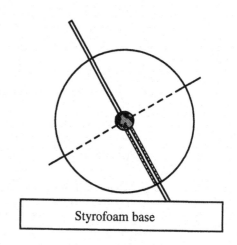

Styrofoam base

> Compare the vertical direction for the observer with your vertical direction.

> Toward which star does the axis of the dowel point?

E. Cut a piece of paper to form a collar that fits around the plastic sphere as shown below. The top of the paper should be level with the center of the earth-bead. The plastic sphere should fit snugly within the collar, but be free to rotate.

The top of the collar represents the horizon for the observer. Mark north, south, east, and west on the paper at the appropriate locations. Also mark azimuths every 10° starting from 0° (north).

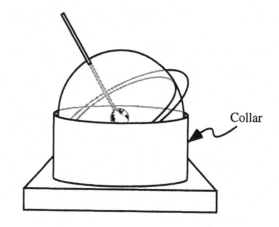

Collar

✔ Check your apparatus with a staff member.

McDermott & P.E.G., U.Wash./*Physics by Inquiry*

The plastic sphere in Experiment 1.2 represents the celestial sphere. Keep in mind, however, that the celestial sphere we imagine as surrounding the earth is much larger than the earth. We should envision the earth-bead as being much smaller than it actually is.

We begin our investigation of the celestial sphere by examining the behavior of stars at various locations in the sky. We then find the current location of the sun on the sphere.

Experiment 1.3

Obtain a small adhesive dot to place on your celestial sphere. The dot will represent a star.

A. Arrange the earth-bead so that your location on the earth is on top. Imagine an observer at your location on the earth-bead. Place the dot on the sphere so that it represents a star directly above the observer.

Hold the dowel so that the earth bead is stationary, and rotate the celestial sphere around the earth.

In which direction should you rotate the sphere so that the star would appear to move in the correct direction across the sky for the observer?

Does the star ever rise and set for the observer? If so, find the approximate directions in which it rises and sets. Express your answer in terms of azimuth. If the star does not set, describe its motion.

B. On what portion of the celestial sphere must you place a star so that for the observer at your location:

- the star never rises above the horizon

- the star never goes below the horizon

- the star is in the sky for essentially half of the day

- the star is in the sky for less than half of the day

- the star is in the sky for more than half of the day

Remove the dot representing the star before continuing.

Experiment 1.4

For this experiment, you will need a recent shadow plot and a small adhesive dot to represent the sun.

A. Find the altitude of the sun at local noon from the shadow plot.

Place the dot representing the sun on your sphere. Choose a location that is consistent with the noon-time altitude of the sun for the day of the shadow plot. Explain how you determined where to place the dot.

B. Hold the dowel so that the earth bead is stationary, and rotate the celestial sphere around the earth.

In which direction should you rotate the sphere so that an observer would see the sun move in the correct direction across the sky?

Estimate the number of hours that the sun was above the horizon on the day the shadow plot was made. Explain your reasoning. (*Hint:* Use the notches on the celestial sphere.)

Estimate the azimuth of the sun at sunrise and at sunset on the day the shadow plot was made. (*Hint:* Use the azimuth scale on the paper collar that you constructed in Exercise 1.2.)

C. Check that your answers in part B are consistent with the shadow plot.

D. Obtain a newspaper for the day you made your shadow plot. Are the times listed in the newspaper for sunrise and sunset consistent with your results in part B? Try to resolve any discrepancies.

E. There are many points on the sphere where you could have placed the dot representing the sun so that it would correspond to the correct noon-time altitude. Identify these locations on the sphere and trace the set of points with your finger.

✔ Check your answers to Experiments 1.3 and 1.4 with a staff member.

The point directly overhead of an observer is called the *zenith.* We can envision a great circle on the celestial sphere that goes from due south on your horizon, through the zenith, then to due north on your horizon. This great circle is called your *local celestial meridian.* The sun, the moon, and the stars are said to *transit* your local celestial meridian at the time they are seen crossing it. We can also imagine a great circle on the celestial sphere directly above the equator on the earth. This circle is called the *celestial equator.*

McDermott & P.E.G., U.Wash./*Physics by Inquiry*

Experiment 1.5

In this experiment, use the sun dot on your celestial sphere from Experiment 1.4.

Imagine an observer at your location on the earth. Rotate the celestial sphere so that it represents local noon for this observer.

A. Trace the local celestial meridian for that observer with your finger.

Point to the part of the sky that corresponds to your own local celestial meridian. Explain how you arrived at your answer.

B. Repeat part A for the zenith and for the celestial equator.

C. Suppose that the observer had seen another object move across the sky on this day along the same path as the sun. Place a second dot on the celestial sphere to represent the location of this object.

Is there more than one possible location on the celestial sphere for this object? If so, describe all the possible locations on the celestial sphere where you could have placed your second dot. If not, explain why not.

D. Now place the second dot at a location on the celestial sphere so that it represents an object located on the observer's zenith at local noon.

Rotate your celestial sphere to show the path that this object takes across the sky. Describe the locations of all points on the celestial sphere that follow the same path across the sky as this object.

In the preceding experiments, you determined the locations of all the points on the celestial sphere that transit your local celestial meridian at the same altitude. These points all lie on a circle around the celestial sphere. They are said to have the same *declination.*

The declination of a point on the celestial sphere is its angle above or below the celestial equator. A point on the celestial equator has a declination of zero. A point located north of the celestial equator has a positive declination; a point south of the celestial equator has a negative declination. All points with the same declination follow the same path across the sky.

Experiment 1.6

A. Use a marker to draw and label lines on your celestial sphere that correspond to the following declinations: 60°, 30°, –30°, and –60°.

Point to the parts of the sky that correspond to the lines of declination that you have just drawn. (Do this for as many of the lines of declination above as possible.) Explain your reasoning.

Some stars are always above the horizon, others never come above the horizon. Use your celestial sphere to determine the range of declinations of stars in each category. Explain your reasoning.

B. Use a *non-permanent* marker to draw a line on your celestial sphere corresponding to your local celestial meridian at a given time.

Do all points on the local celestial meridian have the same declination? Explain.

Does the declination of an object affect the length of time that the object is above the horizon? If so, compare the time that objects with positive, zero, and negative declination are above your horizon.

Would the line that you have drawn for the local celestial meridian also represent your local celestial meridian several hours later? Explain.

Erase the local celestial meridian line before continuing.

C. The number of daylight hours is greater in the summer than in the winter. What does this observation suggest about how the declination of the sun changes during the year? Explain.

What other evidence do you have that the location of the sun changes with respect to the stars?

Experiment 1.7

For this experiment, you will need the shadow plot you used in Experiment 1.4, as well as two other shadow plots. If possible, use shadow plots that were made several months apart.

A. Find the approximate declination of the sun on the day that the shadow plot you used in Experiment 1.4 was made. Explain how you arrived at your answer.

B. Repeat part A for your remaining two shadow plots.

C. Suppose that on three different days during the year, the sun had declinations of (1) 0°, (2) 15°, (3) –15°.

   In each case, find the altitude of the sun at local noon on that particular day. Explain how you arrived at your answers.

D. Summarize the methods that you used in this experiment:

   Describe how you could determine the declination of the sun on a given day if you know its noon-time altitude on that day.

   Describe how you could determine the noon-time altitude of the sun on a given day if you know its declination on that day.

Experiment 1.8

For this experiment, use the dot on your celestial sphere that represents the sun on the day that the shadow plot in Experiment 1.4 was made. All of the questions below refer to that date.

A. Suppose a star transits your local celestial meridian 12 hours after the sun. Determine the location of the star on the celestial sphere with respect to the sun.

   Is there more than one possible location? If so, identify all other possible locations of the star.

   Suppose a star transits your local celestial meridian at the same time as the sun. Determine the location of the star on the celestial sphere with respect to the sun.

   Is there more than one possible location? If so, identify all other possible locations of the star.

B. Determine the location of a star on the celestial sphere that transits your local celestial meridian:

   - 6 hours after the sun
   - 18 hours after the sun

C. Explain how you can use the notches on the perimeter of your sphere to determine how much later than the sun you would see a star transit your local celestial meridian.

✔ Check your answers to Experiments 1.7 and 1.8 with a staff member.

As illustrated in the preceding experiments, many points on the celestial sphere have the same declination. We need another measurement in order to specify uniquely the location of an object on the celestial sphere.

Imagine a line on the celestial sphere extending from the north celestial pole to the south celestial pole. (The celestial poles are the points on the celestial sphere at which the axis of rotation of the earth intersects the celestial sphere.) The stars that you see along this line transit your local celestial meridian at the same time and are said to have the same *right ascension.* Taken together, the declination and the right ascension determine the location of an object on the celestial sphere. Since each line of right ascension transits your local celestial meridian at a different time, we can measure right ascension in units of time: from 0 to 24 hours. In the following section, we introduce the convention for measuring right ascension.

Section 2. Annual motion of the sun and stars: a geocentric model

In the preceding section, we found that the location of the sun changes relative to the stars from day to day. In this section and the next, we develop two models that can help us to account for this observation.

In our first model, we imagine that the sun and the stars all lie on the surface of a very large sphere with the earth at its center. This is called a *geocentric* model and is represented by the celestial sphere that you constructed in the preceding section.

Experiment 2.1

In this experiment, we draw on your observations of the altitude and azimuth of stars at the same time of day.

Choose a star near the celestial equator that you have observed for several months. For each of your observations of the star you will also need a shadow plot for the corresponding week. Identify the name of the star. If you do not know its common name, choose a name for it.

Remove any adhesive dots that are on your sphere.

A. For the first of your star observations:

(1) Place a dot on the sphere to represent the star. Place it so that an observer at your location would see the star at approximately the same altitude and azimuth as you did for your first observation. (Check that the collar on which you have labeled the azimuth is correctly oriented.) Write the name of the star on the dot.

Estimate the declination of the star. Explain your reasoning.

(2) Place a dot on the sphere to represent the sun at the time of your star observation. Use different colors for the sun and star dots. The steps below can serve as a guide to helping you determine where to place the dot for the sun.

   • Determine the declination of the sun for the day of your star observation. Use a shadow plot for that day or that week.

   • Determine the location of the sun at the time of your star observation. Place a dot on the sphere to represent the sun. (*Hint:* At the time of your star observation how many hours had passed since local noon?)

McDermott & P.E.G., U.Wash./*Physics by Inquiry*

- The locations of the sun and star now correspond to the first star observation that you made. Write the date of the observation on the dot representing the sun.

- Check that your locations of the sun and star are correct by starting at local noon for that day and rotating the sphere forward by the number of hours corresponding to your star observation. Is the star in approximately the correct location for your star observation?

B. In plotting your subsequent observations, assume the sun changes location on the celestial sphere, but the stars are fixed to the sphere.

(1) Rotate the sphere so that the star is at the altitude and azimuth corresponding to another of your star observations. (If possible, use an observation that was made about one month after your first observation.)

(2) Place a dot on the sphere to show the location of the sun relative to the star on the day of the second observation. Use the same procedure as part A. Write the date of the observation on the dot representing the sun.

C. Repeat part B for several more of your star observations. Plot enough points that you can tell the approximate shape of the path of the sun across the celestial sphere.

Describe the path of the sun across the celestial sphere.

D. In Experiment 1.6 you discussed how the declination of an object affects the amount of time that object is above your horizon.

Is the path of the sun on the celestial sphere consistent with your knowledge of how the length of the day changes during the year? Explain.

Where on the celestial sphere do you think the sun would be located one year after your first star observation? Explain your reasoning.

✔ Check your answers to this experiment with a staff member.

Do not remove the dots representing the sun from your sphere. You will need them for Experiment 2.2.

The path of the sun on the celestial sphere is a circle called the *ecliptic*. The sun reaches its highest declination on approximately June 21. This occurrence is called the *summer solstice*. The latitude on the earth corresponding to this declination is called the *tropic of Cancer*. (For a discussion of the term *latitude*, refer to *Astronomy by Sight* in Volume I.) The sun reaches its lowest declination on or near December 22, the *winter solstice*. The corresponding latitude on the earth is called the *tropic of Capricorn*.

The sun is located on the celestial equator twice each year. These two events are called the *equinoxes*. One occurs on approximately March 21 and is called the *vernal equinox*. (*Vernal* means "of or related to the spring.") The *autumnal equinox* occurs approximately September 23.

Experiment 2.2

A. Trace out the ecliptic on your celestial sphere with your finger. Identify the location of the sun on the celestial sphere on each of the equinoxes and the solstices.

B. Suppose a star transits your local celestial meridian at the following times on the vernal equinox. Describe the locations on the celestial sphere where the star could be located. Explain your reasoning.

- 6 hours after the sun

- 12 hours after the sun

- 18 hours after the sun

Experiment 2.3

Use the celestial sphere in answering the following questions:

A. Is the sun ever directly overhead at your location? If so, where on the ecliptic is the sun located when it passes directly overhead at local noon? Explain.

B. Are there any regions of the earth where the sun is *never* directly overhead at local noon? If so, which region(s)? Explain.

C. Are there any regions of the earth where the sun is *sometimes* directly overhead at local noon? If so, which region(s)? Explain.

D. Are there any regions of the earth where the sun is *always* directly overhead at local noon? If so, which region(s)? Explain.

Experiment 2.4

Consider a star on your celestial sphere that is located near the ecliptic.

Is that star visible at your location every day during the year or only during a certain part of the year? Explain your reasoning. Draw a diagram to illustrate your answer. (*Hint:* Can a star be above the horizon, yet not be visible? Assume that cloud cover is not an issue.)

In Section 1, we introduced right ascension (R.A.) as a way of helping to determine the position of an object in the sky. We found that if we know both the declination and the right ascension of an object, we know the exact location of the object on the celestial sphere. The choice of the line corresponding to zero right ascension is arbitrary. Once chosen, however, the right ascension of all other objects is then uniquely determined.

The right ascension of the sun on the vernal equinox is defined to be zero. Since the sun changes location on the celestial sphere, its right ascension changes during the year. Lines of right ascension are considered fixed to the celestial sphere. Since we have assumed that the stars are also fixed on the sphere, the right ascensions of the stars do not change. The right ascension of a celestial object is the number of hours after the sun that the object transits your local celestial meridian on the vernal equinox.

Experiment 2.5

A. Use your finger to trace the following lines of right ascension on your celestial sphere: 0 hours, 6 hours, 12 hours, and 18 hours.

Through how many degrees must you rotate the celestial sphere to correspond to one hour of right ascension?

B. Determine the right ascension of the sun on the autumnal equinox, on the summer solstice, and on the winter solstice.

C. By what amount does the right ascension of the sun change each month?

D. What is the right ascension of the star that you used in Experiment 2.1?

✔ Check your answers with a staff member, then draw and label on your celestial sphere the lines of R.A. that you identified in part A.

Experiment 2.6

A. The star Betelgeuse in the constellation of Orion has a right ascension of about 6 hours. The declination of Betelgeuse is about +7°.

Place a dot on your celestial sphere to represent Betelgeuse.

On the vernal equinox, how many hours after the sun transits your local celestial meridian will Betelgeuse transit your local celestial meridian?

On the vernal equinox, how many hours after Betelgeuse transits your local celestial meridian will the sun transit your local celestial meridian?

B. Determine the time of day at which Betelgeuse transits your local celestial meridian on the following dates:

- the autumnal equinox

- the summer solstice

- the winter solstice

- today's date

C. Explain how to use the right ascension of an object to determine the time at which it will transit your local celestial meridian on any given day.

✔ Check your answers with a staff member.

Experiment 2.7

Obtain an earth globe.

A. Find the tropic of Cancer and the tropic of Capricorn on the globe. What are their corresponding latitudes?

In Experiments 2.1 and 2.2 you traced out the ecliptic on your celestial sphere. Are the latitudes of the tropic of Cancer and the tropic of Capricorn on the earth globe consistent with your results in those experiments? Explain.

B. Determine the altitude of the sun at local noon in your city when:

- the sun is directly over the tropic of Cancer

- the sun is directly over the tropic of Capricorn

Draw diagrams to illustrate your answers.

If shadow plots for your location on the summer and winter solstices are available, use them to check your answers above.

Experiment 2.8

A. Describe how you could determine the declination of a star on the basis of observations that you make of that star. Explain your reasoning.

B. Describe how you could determine the right ascension of a star on the basis of observations that you make of that star. Explain your reasoning. (*Hint:* Imagine that you have access to an unmarked celestial sphere.)

Section 3.   Annual motion of the sun and stars: a heliocentric model

In the geocentric model that we developed in the last section, we envision the sun and the stars as moving around the earth. This model reproduces what we observe when we look at the sky: the earth is at rest and the heavens are in motion. We can also account for our observations by considering the sun and stars as fixed while the earth moves. This is called a *heliocentric* model of the sky.

Experiment 3.1

For this experiment you will need a light bulb "sun" at about eye level. Your head will represent the earth.

Put a label on a wall in the room to represent the star that you used in Experiment 2.1.

A. Position yourself so that your back is toward the "sun" and the "star" is straight ahead of you (i.e., so that the star is on your local celestial meridian).

What is the time of day represented by this configuration? Explain.

B. Suppose that it is one month later than the date in part A. How must you stand to represent the same time of day as in part A? (*Hint:* Where should the star be located in your field of view?) Base your answer on your observations of the stars over the past few months.

Are you still facing directly toward the star or toward another location in the sky?

In which direction around the sun did you move? Answer in terms of clockwise or counterclockwise as viewed from above your head.

C. Repeat part B for a date two months after the date in part A.

D. If you were to make observations over the course of a year, the stars would appear to be in the same location in the sky one year later. You already know from the observations you have made that certain stars are visible only during part of the year.

Use these observations to help you draw a diagram that illustrates the motion of the earth over the period of one year. Explain your reasoning.

✔ Check your reasoning with a staff member.

In Experiment 2.1 you found the altitude of the sun at local noon for several days during the year. In the following experiment we incorporate this information into our heliocentric model.

Experiment 3.2

A. Obtain an earth globe and a light bulb "sun." Hold the earth so that the axis of rotation is vertical as shown below.

Sun

Earth

Axis of
rotation

For this arrangement of the sun and earth, make a sketch that shows the altitude of the sun at local noon for an observer at your location.

What is the declination of the sun for this configuration?

What time(s) of year does this configuration represent? Explain.

B. Arrange the light bulb sun and earth globe to represent the summer solstice. Keep the axis of rotation of the earth vertical. Explain why the axis should remain vertical. (*Hint:* Think about where an observer on the earth globe would look for Polaris during the course of a year.)

For this arrangement of the sun and earth, make a sketch that shows the altitude of the sun at local noon for an observer in your city.

What is the declination of the sun for this configuration? Explain how you can tell from your diagram.

C. Repeat part B for the winter solstice.

D. Move the earth around the sun to show the relative orientation of the earth and sun during the passage of one year.

✔ Check your reasoning with a staff member.

Exercise 3.3

In the preceding experiment, you kept the axis of the earth globe vertical. It is possible to represent the motion of the earth around the sun in another way: Imagine that the earth moves in a horizontal circle around the sun.

Hold the earth globe at the same height as the light bulb sun and move it in a horizontal circle around the light bulb sun.

How do you have to orient the earth globe so that the arrangement of the globe and bulb are consistent with your observations of the stars? (*Hint:* Think about where Polaris would be with this orientation so that an observer on the earth globe would see it at the correct location during the course of the year.)

In this model, where would you have to place a label to represent the constellation that you used in Experiment 2.6?

It is often said that the axis of the earth is "tilted." Describe what is meant by this statement. With respect to what is the axis tilted? What is the angle of the earth's "tilt"?

✔ Check your reasoning with a staff member.

Exercise 3.4

Together with three other students in your class, act out the motions of the sun, moon, earth, and a star. Each person should play the role of one of the four objects. Illustrate the geocentric and the heliocentric models that we have developed. In each case, act out enough of the motions so that the daily, monthly, and yearly motions are evident.

Exercise 3.5

Discuss with your partners the geocentric and the heliocentric models for the sky developed in this section. Are you able to account for your observations with both models? Do these observations provide a basis for you to prefer one model over the other?

Section 4. The seasons

On most parts of the earth, there are noticeable changes in light from the sun and temperature from month to month. In this section, we account for these seasonal changes on the basis of our observations of the sun.

Experiment 4.1

A. Place a small adhesive dot on your celestial sphere to represent the sun on the vernal equinox.

At your location on the earth, for how many hours is the sun above the horizon on the vernal equinox?

Choose a location on the earth that has the same latitude as your location but that is on the other side of the equator.

At that location on the earth, for how many hours is the sun above the horizon on the vernal equinox? Check your answer using your celestial sphere.

B. Repeat part A for each of the following dates: the autumnal equinox, the summer solstice, and the winter solstice.

How does the length of time that the sun is above the horizon for your location change during the year?

The previous experiment illustrates one aspect of how the sun's motion on the ecliptic contributes to the amount of light that reaches a location on earth. Another aspect is illustrated in the following experiment.

Experiment 4.2

A. Turn on a flashlight in a darkened room. Hold an index card or a small piece of paper in a vertical position about one meter away from the flashlight, as shown in the side-view diagram at right. The paper should be small enough so that the paper is entirely lit when it is in this position.

Small piece
of paper

Flashlight

McDermott & P.E.G., U.Wash./*Physics by Inquiry*

Now rotate the paper gradually until it is horizontal. Describe your observations.

How does the total amount of light falling on the entire piece of paper change when the paper is rotated? Explain your reasoning.

B. The diagram below shows light from the sun incident on the earth. For this arrangement of the earth and sun:

What is the declination of the sun? Explain how you can tell from the diagram.

What time(s) of year does this diagram represent? Explain your reasoning.

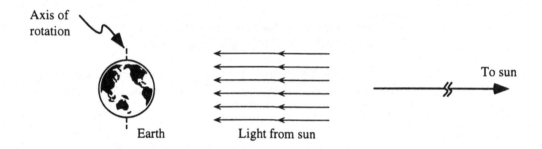

C. Consider three small farms on the earth, all the same size. The first is on the equator, the second is at a latitude of 30° north of the equator, the third is 30° south of the equator.

(1) On the day represented by the diagram in part B:

For any given minute that the sun is in the sky, which farm receives the most light from the sun? Explain.

(2) Sketch a diagram similar to the one in part B to show the earth and light from the sun on the summer solstice.

For any given minute that the sun is in the sky on the summer solstice, which farm receives the most light from the sun? Explain.

(3) Sketch another diagram to show the earth and light from the sun on the winter solstice.

For any given minute that the sun is in the sky on the winter solstice, which farm receives the most light from the sun? Explain.

Exercise 4.3

Use your results from Experiments 4.1 and 4.2 in answering the following questions.

Consider a location on the northern hemisphere.

A. Is the total amount of light received from the sun at this location greater during the time that the sun is (1) north of the celestial equator or (2) south of the celestial equator? Explain your reasoning.

B. On the basis of your answer above, would you expect more heat to be transferred from the sun to this location during the time that the sun is (1) north of the celestial equator or (2) south of the celestial equator? Explain your reasoning.

C. Repeat this exercise for a location on the southern hemisphere.

Exercise 4.4

On the basis of your results from the preceding experiments:

A. Discuss the factors that are responsible for the seasons in the northern and southern hemispheres. Explain your reasoning.

B. Compare the seasons in the northern hemisphere and in the southern hemisphere.

✔ Discuss this exercise with a staff member.

Exercise 4.5

A student makes the following statement:

> *"The seasons occur because the distance between the earth and sun changes during the year. We are farther from the sun in the winter and closer to it in the summer."*

On the basis of your results so far, give as many arguments as you can against this student's belief. Do you have any observations that suggest the distance between the earth and the sun changes during the year? Explain.

     McDermott & P.E.G., U.Wash./ *Physics by Inquiry*

In the regions of the earth near the poles, the change from winter to summer can be quite dramatic. The number of daylight hours varies enormously. In the following experiment we investigate this effect.

Experiment 4.6

For this experiment you will need an earth globe and a light bulb to represent the sun.

A. Arrange the globe and light bulb to represent the summer solstice. Rotate the globe through one entire day.

Describe the motion of the sun in the sky for an observer located:

- at the north pole

- at a latitude a few degrees south of the north pole

- at a latitude of 45° north of the equator

- at the equator

- at a latitude of 45° south of the equator

- at a latitude a few degrees north of the south pole

- at the south pole

B. Sketch a diagram to show light from the sun incident on the earth on the summer solstice.

(1) On the summer solstice there is a region of the earth for which the sun remains above the horizon all day and does not set.

Indicate this region on your diagram. Find the range of latitudes of this region (e.g., from ___ to ___ degrees north of the equator). Use geometry and your knowledge of the latitude of the tropic of Cancer to answer this question.

(2) On the summer solstice there is a region of the earth for which the sun remains below the horizon all day and does not rise.

Indicate this region on your diagram. Find the range of latitudes of this region. Explain your reasoning.

C. For any region of the earth, does the sun remain below the horizon or stay above the horizon on any of the following dates? Explain.

- the autumnal equinox

- a day halfway between the autumnal equinox and the winter solstice

- the winter solstice

- a day halfway between the winter solstice and the vernal equinox

There are two regions of the earth in which the sun does not rise above the horizon for at least one day of the year, one in the northern hemisphere, and the other in the southern hemisphere. The southern border of the region in the northern hemisphere is marked by a line of latitude called the *Arctic circle*. The northern boundary of the region in the southern hemisphere is called the *Antarctic circle*.

Exercise 4.7

Obtain an earth globe and check your predictions from part B of Experiment 4.6.

✔ Check your reasoning with a staff member.

Experiment 4.8

For this experiment you will need an index card, a thumbtack, some double-sided tape, an earth globe, and a bright light bulb.

Cut the index cards into small squares, each approximately 5 cm on a side. Push the thumbtack through the center of one of the small squares to form a miniature gnomon. Put a small piece of double-sided tape on the back of the gnomon.

We will use these miniature gnomons to make shadow plots for various locations on the earth and for various times of the year.

A. Place a gnomon on the globe at a position that corresponds to your location on the earth. Place the light bulb so it is at the same height as the center of the globe.

McDermott & P.E.G., U.Wash./ *Physics by Inquiry*

B. Make shadow plots for your location on the earth that correspond to several different days during the year. Write the time of year and location on each plot. (Do not try to write the time of day for each of your marks.)

Are your miniature shadow plots consistent with the real plots that you have made over the past few months? If not, try to account for any discrepancies.

C. Make shadow plots for some other locations on the globe. For each location make several plots, each for a different day of the year. Include the following plots:

- the equator on one of the equinoxes

- the north pole on one of the equinoxes

- the north pole on the summer solstice

- the south pole on the summer solstice

Section 5. The planets

While most of the stars in the sky remain fixed with respect to one another, a few appear to "wander." These are called *planets* (the Greek word for *wanderers*). In this section, we extend our models of the sky to account for our observations of the planets.

Begin this section after you have observed the planets for a period extending over at least several months.

Exercise 5.1

Discuss with your partner some of the facts that you know about the planets. Make a list of all the items you mention.

A. Separate your list into two categories: those items for which you have direct evidence and those that you know because you were told them or you read about them.

In the case of those items for which you believe you have direct evidence, describe your evidence and explain how it is relevant.

B. Which of the facts on your list do you think you could have come to know by observing the planets yourself (without materials or equipment to which you do not have ready access)? In each case, describe the observations you would need to make.

Exercise 5.2

A. Choose a planet that you have observed. In the center of a large sheet of paper, sketch a diagram showing the background stars near that planet. On this diagram, for each of your observations of that planet:

(1) Mark the location of the planet. Record the date and time next to that location.

(2) If the sun was in the vicinity of the planet during any of your observations, record its location too (along with the date).

This diagram will be your *observation chart* for this particular planet.

B. Repeat part A for each of the other planets that you have observed.

    McDermott & P.E.G., U.Wash./ *Physics by Inquiry*

C. Examine your observation charts.

Try to find patterns in the motion of each planet in the sky.

Try to find any similarities between the behavior of the different planets.

Discuss with your partner any features of the data that you find interesting or noteworthy.

D. Base your answers to the following questions on your observations of the planets.

(1) During the course of a single day:

How does the apparent motion of the planets compare to the apparent motion of the stars and the sun?

(2) During the entire period of your observations:

How is the motion of the planets across the sky similar to or different from that of the stars?

How is the motion of the planets across the sky similar to or different from that of the sun?

✔ Check your answers with a staff member.

Motion of the planets in a geocentric model

In a geocentric model, we think of the stars as fixed on the celestial sphere. The sun moves slowly around this sphere once each year. In the following experiments and exercises, we interpret our observations of the motion of the planets in geocentric terms.

Experiment 5.3

In this experiment, we consider the motion of Venus. For simplicity, we only consider the general features of this motion.

Obtain a set of three balls to represent the earth, the sun, and Venus. In this experiment, we will think of the stars, sun, and planets as moving around the earth.

A. Move the balls so that they reproduce the motion of the sun and Venus as viewed from the earth during the course of one day. (*Hint:* Do the relative locations of Venus and the sun change substantially in a day?)

B. Arrange the three balls to represent the relative locations of the sun, Venus, and the earth at local noon on the date of your first observation of Venus.

Now arrange the balls to represent the relative locations of the three objects at local noon on the date of your next observation of Venus. (*Hint:* Consider whether the location of the ball that represents the sun would be different from what it was before.)

Does the observed location of Venus change with respect to that of the sun during the entire period of your observations? If so how?

Does the observed location of the sun change with respect to that of the background stars during the entire period of your observations? If so how?

C. Repeat part B for the remainder of your observations of Venus. For each day that you observed Venus, arrange the balls to represent the relative locations of the sun, Venus, and the earth at local noon on that day.

Describe how it is possible to account for your observations by thinking of Venus as moving in a circle around the sun.

Describe at least two other ways in which Venus might move that would also account for your observations. (*Hint:* Suppose, for example, that Venus either moves in a circle around some other location or that it moves in a path that is not circular.)

Is it possible to think of Venus as moving in a circular path around the earth? Explain.

Exercise 5.4

In this experiment, we consider the motion of Mars. As in the preceding experiment, we consider only general features of the motion and regard the earth as fixed. You will need your observation chart for Mars from Exercise 5.2.

A. Describe the motion of the sun, Mars, and the background stars that appear near Mars during the course of one day. (*Hint:* Does the observed location of Mars relative to the background stars change substantially in one day?)

B. Describe how the location of Mars in the sky changes with respect to the background stars over the entire period of your observations.

Does Mars appear to move eastward or westward with respect to the stars? Is this motion always the same?

We can imagine that Mars moves in a circle around an unseen object that moves around the celestial sphere in the same direction as the sun. Explain how you can use this model to account for your observations. Support your answer by drawing a diagram that shows how both Mars and the unseen object would move around the celestial sphere.

Describe at least one other way that Mars might move that would account for your observations.

Is it possible that Mars is closer to the sun than the earth? Explain how you can tell from your observations.

C. When Mars appears to move westward with respect to the stars, it is said to be undergoing *retrograde motion*.

Describe how the model in part B can be used to account for the retrograde motion of Mars. Include a diagram with your explanation that clearly illustrates the retrograde motion of Mars.

✔ Check your reasoning with a staff member.

Venus and Mars are not the only planets that are visible with the naked eye. You may already have observed Jupiter and Saturn. Other planets, including Mercury, Uranus, Neptune, and Pluto, are most readily observed (if at all) with the aid of optical instruments. Together with the sun and the earth, all of these planets are said to be part of the *solar system*.

Exercise 5.5

If you have made observations of any planets besides Venus and Mars, answer the following questions for each of those planets:

Is the motion of the planet in the sky more like that of Venus or Mars? Explain your reasoning.

Show how you can account for your observations in the geocentric model. (*Hint:* Should you think of the planet as moving in a circular path around the sun, or should you think of it as moving in a circular path around an unseen body that moves in the same direction along the celestial sphere as the sun?)

Motion of the planets in a heliocentric model

Although they were quite complicated, early geocentric models of the universe could account for the observed motion of the planets. We now investigate how we can account for our observations with a heliocentric model. In this model, we think of the sun and stars as fixed, and we think of the earth as both rotating once each day around its axis and moving once each year around the sun.

Experiment 5.6

In this experiment, we consider the motion of Venus. Set up a light bulb in the center of the room to represent the sun. Place labels around the perimeter of the room to represent different fixed stars. Decide with your partner who will play the role of Venus and who will play the role of the earth.

A. You and your partner should stand so that your positions relative to the bulb are consistent with one of your early observations of Venus.

B. Now you and your partner should move so that your positions are consistent with your next observation of Venus.

C. Repeat part B for all of your observations of Venus.

D. Show that it is possible to account for your observations of Venus by thinking of Venus as moving in a circle around the sun.

Is Venus closer to the sun or farther from the sun than the earth? Explain how you can tell from your observations.

E. Describe at least two other ways that Venus might move that would account for your observations.

Experiment 5.7

In this experiment, we consider the motion of Mars in the heliocentric model. You will need your observation chart for Mars from Exercise 5.2.

Set up a light bulb to represent the sun. Place labels around the perimeter of the room to represent different background stars. Decide with your partner who will play the role of Mars and who will play the role of the earth.

A. You and your partner should stand so that your positions relative to the bulb are consistent with one of your early observations of Mars.

B. Now you and your partner should move so that your positions are consistent with your next observation of Mars. Be sure that an observer on earth would see Mars change location with respect to the background stars in a way that is consistent with your observations.

C. Repeat part B for your remaining observations of Mars.

D. Show that it is possible to account for your observations by thinking of Mars as moving in a circle around the sun. (*Hint:* In order to account for your observations, should Mars or the earth take longer to go once around the sun?)

Is Mars closer to the sun or farther from the sun than the earth? Explain how you can tell from your observations.

Describe at least one other way that Mars might move that would account for your observations.

E. If you have not done so already, use your physical model to account for the retrograde motion of Mars.

✔ Check your reasoning for this experiment and the preceding experiment with a staff member.

Exercise 5.8

For each of the planets that you have observed:

Show that it is possible to account for your observations in the heliocentric model by thinking of the planet as moving in a circle around the sun.

Determine if the planet is closer to the sun or farther from the sun than the earth.

Exercise 5.9

We have found that two different models can account for our observations of the motion of the planets. Do these observations provide a basis for you to prefer one model over the other?

Discuss this question with other students and the staff.

Supplementary problems for *Astronomy by Sight: the earth and the solar system*

Problem 1.1

For this problem, you will need a recent shadow plot, your celestial sphere, and a small adhesive dot to represent the sun.

A. Find the altitude of the sun at local noon for the day you made your shadow plot. Place the dot representing the sun on your sphere at an appropriate location. Explain how you decided to place the dot where you did.

B. Use your celestial sphere to help you answer the following questions about the day that you made your shadow plot.

Choose a distant city on the earth with the same latitude as your own.

When it was local noon in that city, what time was it in your city? Explain.

What altitude did the sun have at local noon in that city? Explain.

Choose a distant city on the earth with the same longitude as your own.

When it was local noon in that city, what time was it in your city? Explain.

What altitude did the sun have at local noon in that city? Explain.

Problem 1.2

Two students are making observations of the sky one night, and are talking on the telephone. The first student describes a star that is located on his local celestial meridian, 10° north of his zenith. The second student, who is in Seattle, observes the same star (at the same instant) exactly on her zenith. Determine the location of the first student. Explain.

Problem 2.1

Choose two cities in addition to your own. Each city should have a very different latitude from the others. Use your celestial sphere to help you answer the following questions.

A. For each city, determine for how many hours the sun is above the horizon on each of the equinoxes. Explain how you determined your answers.

Do you think your answers would have been different if you had chosen different cities? Explain.

B. Repeat part A for the summer solstice and the winter solstice.

McDermott & P.E.G., U.Wash./*Physics by Inquiry*

Problem 2.2

Two students are making observations of the sky one night, and are talking on the telephone. The first student, who is in Seattle, describes a star transiting his local celestial meridian at an altitude of 25°. The second student says that the same star transited her local celestial meridian 2 hours before at the same altitude. Determine the location of the second student. Explain.

Problem 2.3

One night at 10:00 P.M. in Seattle, a student observes a star that is due north and at an altitude of 10°. Answer the following questions using your geocentric model for the sky. Explain your reasoning in each case.

A. The student observes the same star one hour later.

    (1) Will the altitude of the star increase, decrease, or stay the same?

    (2) Will the star move eastward, westward, or stay at the same location in the sky?

B. At 10:00 P.M. two weeks later, the student looks into the sky from the same location as before. In which direction should the student look to see the star?

C. At 10:00 P.M. six months later, the student is again at the same location. Will the star be above the student's horizon? If so, determine the altitude and azimuth of the star.

Problem 2.4

At midnight your time, in late February, Matthew calls you from some unknown location of the earth and tells you that he sees the constellation Orion due north. He says the center of the constellation has an altitude of about 30°. What is Matthew's approximate location on the earth?

Problem 2.5

As part of your observations for this module, you observed that the moon changes its location with respect to the background stars from day to day. Describe a procedure by which you could use observations of the moon and the stars to find the path of the moon on the celestial sphere. Explain your reasoning.

Problem 2.6

A. Suppose you measure the altitude of a star to be 45° and its azimuth to be 315° at 12:01 A.M. on February 3. Determine the approximate right ascension and declination of the star. Explain how you arrived at your answer.

B. Suppose that another observer located a few hundred miles to your south observed a star at the same altitude and azimuth at the same date and time as you did. Was it the same star that you observed? If not, compare its right ascension and declination with the star in part A. Explain your reasoning.

Problem 2.7

Do this problem after you have completed at least through Experiment 2.6 of this module.

Below is a table listing the names, right ascensions, and declinations of several bright stars.

Star name	Right ascension	Declination
Aldebaran	4 hours 35 minutes	17°
Alpharatz	0h 08m	29°
Antares	16h 29m	–26°
Arcturus	14h 15m	19°
Canopus	6h 23m	–52°
Denebola	11h 49m	15°
Dubhe	11h 03m	62°
Rigel	5h 14m	–8°
Sirius	6h 45m	–16°
Vega	18h 36m	39°

A. For each star of the stars listed above:

(1) Place a dot on your celestial sphere that represents the location of the star. Label the dot with the name of the star.

(2) Determine the approximate date that the star transits your local celestial meridian at midnight. Explain your reasoning.

(3) Determine the approximate time that the star transits your local celestial meridian on today's date. Explain your reasoning.

B. Which of these stars would you expect to see transit your local celestial meridian with the highest altitude?

C. Which of these stars would you expect to see at midnight tonight? For each star that is above the horizon at that time, predict the altitude and azimuth of each star.

McDermott & P.E.G., U.Wash./*Physics by Inquiry*

Problem 2.8

A star crosses your local celestial meridian with an altitude of 25° and an azimuth of 180° at 3:00 AM on August 7.

Determine the approximate right ascension and declination of the star. Explain how you arrived at your answer.

Problem 3.1

As part of your moon observations you should have recorded the stars that appear near the moon.

On the basis of your observations, is the moon surrounded by the same stars each time it is in a given phase? (For example, is the full moon always near a certain star?) Explain how your answer is consistent with your physical model.

Problem 3.2

We have observed that the sun moves with respect to the stars during the year. We have also observed that the location of Polaris in the sky does not change significantly. Use these observations to answer the following questions.

 Sun

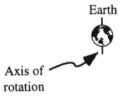 Earth

Axis of rotation

A. The diagram above illustrates the relative locations of the earth and the sun on the vernal equinox. On the diagram, draw a line that points in the direction of Polaris for an observer at your location at midnight on the vernal equinox.

B. Consider the heliocentric model. On the same diagram above, draw the location of the earth six months later (on the autumnal equinox). Draw a line that points in the direction of Polaris for an observer at your location on the autumnal equinox.

If we think of Polaris as not moving, what does your figure suggest about how the distance between the earth and Polaris compares to the distance between the earth and the sun? Explain your reasoning.

C. Consider a modification of the geocentric model, in which the stars (but not the sun) reside *on* the celestial sphere at some large, fixed distance from the earth. Think of the sun as located at some *different* distance from the earth.

If the distance between Polaris and the sun does not change very much, how does the distance between the earth and Polaris compare to the distance between the earth and the sun? Explain your reasoning.

McDermott & P.E.G., U.Wash./*Physics by Inquiry*

Problem 3.3

In Volume I of *Astronomy by Sight,* you found that the moon takes about 30 days to go through one complete cycle of phases. (This is called the synodic period.) You may have also found that the moon takes about 27 days to return to the same position with respect to the background stars. (This is called the sidereal period.)

How can you account for this difference in time using your physical model? Use sketches and an explanation to support your argument.

Problem 4.1

A student makes the following statement:

"The city of Quito, Ecuador, is located on the equator. So, the night there lasts as long as the day throughout the year. This must also mean that every day the sun rises due east at the same time and sets due west at the same time."

Do you agree or disagree with the above statement? Explain your reasoning.

Problem 4.2

We can consider the earth as consisting of five regions separated by the Arctic circle, the tropic of Cancer, the tropic of Capricorn, and the Antarctic circle. Describe the motion of the sun throughout the year as viewed by an observer located in each region.

Problem 4.3

A. Based on observations you have made at your location, does the warmest part of the year come before, after, or on the longest day of the year?

B. Suppose a beaker of water, initially at room temperature, is put over a gas flame. Would the temperature of the water start to rise, fall, or remain constant? Explain.

Long before the water starts to boil, you turn down the flame slightly. Would you expect the temperature of the water to continue to rise, to start falling, or to remain constant? (*Hint:* Is heat still being transferred to the water in this case?)

McDermott & P.E.G., U.Wash./*Physics by Inquiry*

C. Two students who in a city at 40° north latitude discuss the seasons.

Student 1:   *"I know that the greatest daily amount of heat that our town receives from the sun is on the summer solstice. So, that is the warmest day of the year."*

Student 2:   *"I agree that we receive the greatest daily amount of heat from the sun at the summer solstice. However, the day after the summer solstice, we receive almost as much heat as we do on the summer solstice. So, the earth, water, and air are still warming up, and the warmest day of the year could come a month or two later."*

Do you agree with student 1, student 2, or neither? Explain your reasoning.

Problem 4.4

Use the shadow plots that you have made over several months as the basis for your answers below. (If necessary, refer to *Astronomy by Sight* in Volume I.)

A. Plot a graph of the altitude of the sun at your local noon versus time.

Does the altitude of the sun at local noon change appreciably:
- a few days before and after an equinox?
- a few days before and after a solstice?

B. Do sunrise and sunset times change appreciably:
- a few days before and after an equinox?
- a few days before and after a solstice?

C. Are your answers to parts A and B consistent with the meaning of the word solstice, derived from the Latin *solstitium,* "the standing of the sun?"

Problem 5.1

Below is a top view diagram of the earth, Mars, and the sun in the heliocentric model. The diagram shows several locations of the earth and Mars at intervals one month apart.

A. Which path represents the orbit of the earth, and which represents the orbit of Mars? In which direction is each planet traveling? Explain your reasoning.

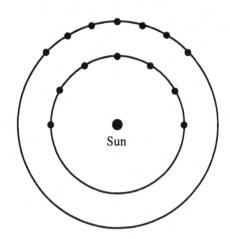

B. Copy the diagram on the bottom of a separate sheet of paper. For each month shown on your diagram, draw a straight line that represents the line of sight for an astronomer on the earth who is observing Mars. From the lines of sight you have drawn, determine how Mars has appeared to change location with respect to the background stars. Explain your reasoning.

C. During which of the months shown, if any, does Mars appear to move:

- *eastward* with respect to the background stars? Explain how you can tell from your diagram.

- *westward* with respect to the background stars (i.e., undergo retrograde motion)? Explain your reasoning.

Problem 5.2

According to observations of Venus, the angle between Venus and the sun changes in a regular pattern.

A. Draw a top view diagram showing the earth, Venus, and the sun in the heliocentric model. Make your diagram consistent with one of your observations. Include in your diagram the orbits of earth and Venus (assume they are circular about the sun).

B. From the location of the earth that you chose, draw one line through Venus and one line through the sun. Identify the angle between Venus and the sun on your diagram for this configuration.

C. Indicate on your diagram the location(s) of Venus for which the angle between Venus and the sun is maximum. Explain your reasoning.

D. Observations show that the angle between Venus and the center of the sun reaches a maximum of 46°.

Using similar triangles, determine the distance between Venus and the sun in terms of the distance between the earth and the sun. (You may find it useful to draw additional lines on your diagram.)

Index

References to POM, H&T, L&C, Mag, AbS 325–379 and the Appendices are in italics and can be found in Volume I of *Physics by Inquiry*.

CPSIA information can be obtained
at www.ICGtesting.com
Printed in the USA
BVOW04s1226270817

492911BV00005B/7/P